PERGAMON INTERNA⌐
of Science, Technology, Engine
The 1000-volume original paperbac⌐
industrial training and the e⌐., ⌐, ⌐.⌐.⌐ ⌐. ⌐⌐⌐⌐⌐
Publisher: Robert Maxwell, M.C.

PRACTICAL WIRING

VOLUME 1

in S.I. Units

ENGINEERING SCIENCE & TECHNOLOGY

CRANE, P. W.
Electronics for Technicians—Worked Examples in Basic Electronics

HANCOCK, N. N.
Matrix Analysis of Electrical Machinery   2nd Edition

HARRIS, D. J. & ROBSON, P. N.
The Physical Basis of Electronics

HINDMARSH, J.
Electrical Machines and their Applications   2nd Edition

LAITHWAITE, E. R.
Exciting Electrical Machines

MYATT, L. J.
Symmetrical Components

RODDY, D.
Introduction to Micro-electronics

WHITFIELD, J. F.
Electrical Installations and Regulations—Electrical Installations Technology

The terms of our inspection copy service apply to all
the above books. A complete catalogue of all books
in the Pergamon International Library is available
on request.
The Publisher will be pleased to receive suggestions
for revised editions and new titles.

PRACTICAL WIRING

VOLUME 1
in S.I. Units

BY

HENRY A. MILLER C.G.I.A., C. Eng., M.I.E.E., F.R.S.A.
Formerly Course Organizer and Lecturer-in-Charge,
Bromley College of Technology

PERGAMON PRESS

OXFORD · NEW YORK · TORONTO
SYDNEY · PARIS · BRAUNSCHWEIG

U. K.	Pergamon Press Ltd., Headington Hill Hall, Oxford OX3 0BW, England
U. S. A.	Pergamon Press Inc., Maxwell House, Fairview Park, Elmsford, New York 10523, U.S.A.
CANADA	Pergamon of Canada, Ltd., 207 Queen's Quay West, Toronto 1, Canada
AUSTRALIA	Pergamon Press (Aust.) Pty. Ltd., 19a Boundary Street, Rushcutters Bay, N.S.W. 2011, Australia
FRANCE	Pergamon Press SARL, 24 rue des Ecoles, 75240 Paris, Cedex 05, France
WEST GERMANY	Pergamon Press GmbH, D-3300 Braunschweig, Postfach 2923, Burgplatz 1, West Germany

Copyright © Henry A. Miller 1969

First edition 1969
Second edition 1975
Library of Congress Catalog Card No. 68-57882

Printed in Great Britain by A. Wheaton & Company, Exeter
ISBN 0 08 019754 X

Contents

CONTENTS

Preface

ELECTRIC wiring is a craft. *The Concise Oxford Dictionary* defines the word "craft" as: "skill; cunning, deceit, art, trade...". The electrician, or electrical craftsman, needs to practise all of these things: skill in the use of tools, cunning in planning and in overcoming difficulties, deceit in concealing wiring wherever possible, art in applying scientific principles, and trade in the sense of earning a reputation for always giving value for money.

This book consists largely of hints to help the would-be craftsman. It is designed to cover the First Year Craft Practice syllabus of the City and Guilds of London Institute A Course in Electrical Installation Work, and similar courses.

H. A. MILLER

Preface to Second Edition

Metrication and the adoption of the International System (SI) has involved a certain amount of revision of this book. The system is based on the six basic units of: length (metre), mass (kilogramme), time (second), electric current (ampere), thermodynamic temperature (Kelvin) and luminous intensity (candela).

Although the book was originally planned to cover the practical content of Electrical Installation Course A, it is also suitable for students taking relevant portions of the City and Guilds of London Institute Overseas Courses 833 (Basic Engineering Trades), 836 (Electrical Installation Practice) and the Council of Technical Examining Boards Course 200 in Basic Engineering Craft Studies Part 1 (Electrical Bias).

H. A. Miller

Hand Tools — Their Use and Care

Use tools only for the jobs for which they are intended.

ONE of the first essentials in wiring is to know the correct way to use and care for the necessary tools. Following are the main items concerning some of the common hand tools used in connection with sheathed wiring (tough-rubber- or PVC-sheathed).

Knives and Cable Strippers

Tools used for cable stripping vary. Pocket knives, cobbler's knives, and trimming knives are typical (Fig. 1.1). In all three cases

1

it is necessary to keep the blade sharp by the occasional use of an oilstone; trimming knives have the advantage that worn blades can easily be replaced.

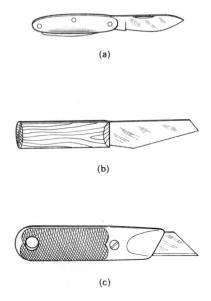

(a)

(b)

(c)

FIG. 1.1. Knives: (a) pocket, (b) cobbler's, (c) trimming.

When stripping tough-rubber- or PVC-sheathed cables it is usual to first ring the sheath at the desired point, making sure not to penetrate right through. After cutting or parting lengthways between the cores, it should be possible to remove the sheath cleanly up to the ringing.

In removing the insulation from a cable core, great care must be taken to avoid nicking the conductor. The knife should therefore be held at an acute angle.

There are a number of cable insulation stripping tools on the market, some good — some not so good. An efficient and well-adjusted stripping tool is particularly useful for stripping PVC-insulated cables in cold weather (Fig. 1.2).

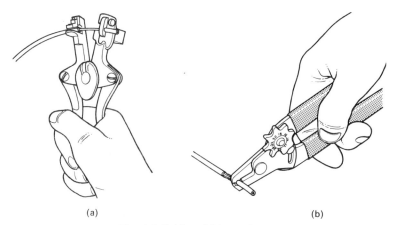

(a)   (b)

FIG. 1.2. Cable-stripping tools.

Hammers

A hammer is used to deliver blows to pins and nails, or to other tools such as cold chisels. As far as electrical work is concerned, the hammer normally consists of a steel head mounted on a wood handle, generally hickory. The head, which is wedged onto the handle, has two working surfaces, known as the face and the pein.

Hammers are classified both by the weight and the type of head. Usual weights vary from 0.1 to 1.50 kg. The most popular type for electricians is the ball pein.

The hammer should be held firmly at the *end* of the handle so that the head is in line with the direction of the blow.

Electricians sometimes mark their hammer handles in cm. This is particularly useful when using the hammer to fix something at specified intervals. For fixing clips or saddles when installing sheathed cables on woodwork, brass pins are often used. Types of nails used in electrical work include: wire, oval, lath, clout, screw, cut clasp (Fig. 1.3).

Screwdrivers

The purpose of a screwdriver is to drive in or unscrew screws. It consists of a hardened and tempered steel or alloy steel blade

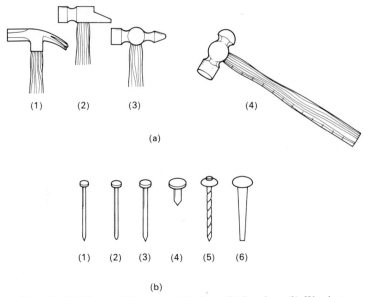

(a)

(b)

FIG. 1.3. (a) Types of hammer: (1) claw, (2) London, (3) Warrington, (4) electrician's ball pein hammer with handle marked in cm. (b) Types of nails: (1) wire, (2) oval, (3) lath, (4) clout, (5) screw, (6) cut clasp.

with a handle of wood or, in the case of electrician's terminal screw-drivers, of moulded insulating material. The end of the blade, known as the tip, is shaped to fit into a screw-head. Ordinary screwdrivers designed for slotted screw-heads may have parallel-sided tips or flared tips. Terminal screwdrivers usually have parallel-sided tips.

The *Phillips* type of screwdriver has a round tip terminating in four flanges designed to fit into a screw-head with crossed recesses. This type of screw is often used in modern electrical appliances, especially those assembled automatically by power-driven screw-drivers. Phillips screwdrivers with specially hardened tips are some-times used for driving in self-tapping screws which cut their own threads.

It is possible to buy ratchet screwdrivers. These enable the handle to be turned clockwise and anticlockwise alternately while turning the blade only in one direction.

4

Use tools only for the jobs for which they are intended. Screws should always be driven in by screwdrivers; *never* by hammers. Screwdrivers are made in several sizes, and it is important to use one with a wide enough tip, otherwise the screw-head may be damaged (Fig. 1.4).

If the tip of an ordinary type of screwdriver becomes worn or damaged, it can be reground on a grindstone.

Screwdrivers are obtainable with removable bits, so that the insert bit can be replaced when worn. In the *Yankee* type, the bit

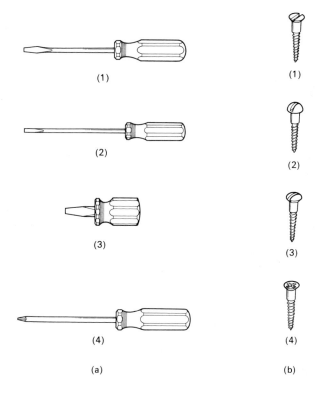

FIG. 1.4. (a) Screwdrivers: (1) flared tip, (2) parallel-sided tip, (3) stub, (4) Phillips. (b) Screw-heads: (1) countersunk, (2) round head, (3) raised head, (4) Phillips.

5

is held in an adaptor by a retaining spring and the float between bit and adaptor compensates for any slight misalignment of the driving head and the screw.

Common types of screw-heads are: countersunk, round head, and raised head.

Bradawls

A bradawl, sometimes called a sprig-bit, is used to make a pilot hole for a wood screw. It comprises a handle of beech or ash with a blade similar to that of a small-size screwdriver. Bradawls are, in fact, also used to turn very tiny terminal screws. The tip may be protected when not in use by a cork. Occasionally a gimlet is used for boring wood (Fig. 1.5).

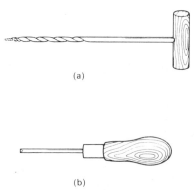

(a)

(b)

FIG. 1.5. (a) Gimlet, (b) bradawl.

Pliers

Pliers are principally used to hold things and, in some cases, to cut wire. Electrician's pliers generally have insulated handles. Many types are available, but those commonly used in electrical work are: side-cutting pliers, snipe-nosed pliers, gas pliers, diagonal-cutting nippers (Fig. 1.6). They are in two or three standard sizes.

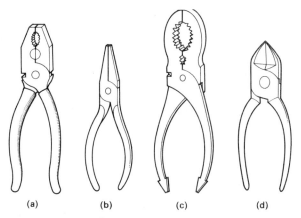

(a)   (b)   (c)   (d)

FIG. 1.6. Pliers: (a) electrician's insulated, (b) snipe-nosed, (c) gas, (d) nippers.

Side-cutting pliers with insulated handles are suitable for general work. They have a square nose with tapered jaws serrated inside for gripping, and wire-cutting edges are provided both in the jaws and at each side of the hinge.

Long snipe-nosed pliers have thin jaws which can be used for making loops at the ends of wires and are also suitable for gripping in confined spaces.

Gas pliers have jaws designed to grip conduits and cable lugs. The jaws are heavily serrated and when gripping care must be taken not to damage the surface of the object. Wire-cutting edges are provided at one side of the hinge.

Diagonal-cutting nippers do not have jaws for gripping. They are used for cutting wire, particularly in places where it would be difficult or impossible to use side-cutting pliers (for example, when removing surplus wire at a terminal).

Wood Saws

The electrician's tool kit usually includes a tenon saw and a padsaw (Fig. 1.7).

Tenon Saws are made in standard sizes.
Two main types are: brass back with closed handle, and steel

7

back with open handle. They are commonly used to cut floorboards which have been prised prior to lifting. When not in use, the teeth of the saw should be protected by a guard consisting of a strip of wood with a groove in which the teeth can be inserted. The guard can be held in place by thick rubber bands.

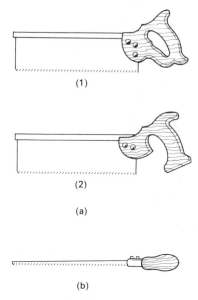

(1)

(2)

(a)

(b)

FIG. 1.7. (a) Tenon saws: (1) closed handle, (2) open handle. (b) Padsaw.

It is possible to sharpen and set tenon saws using a plier-type of saw-set, but those without the necessary knowledge and experience would be well advised to have this done expertly at a tool store which caters for this service.

A padsaw has a thin, detachable blade of cast steel which is secured in a wood handle by screws. Blades are obtainable in lengths of from 200 to 300 mm. The cut is often started at a drilled hole.

Plaster Saws

Sometimes, a plaster saw (Fig. 1.8) is included among the tools. This is designed to cut a groove in plaster when wiring has to be sunk in walls at switch drops, etc. The parallel blades are adjustable, so that narrow or wide slots up to 50 mm may be cut.

A wall-chasing machine is also available for this job.

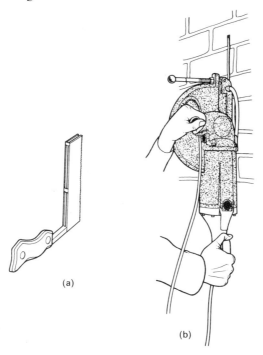

(a)

(b)

FIG. 1.8. (a) Plaster saw, (b) wall-chasing machine.

Wood Chisels

In spite of the rapid replacement of wood by moulded plastics in electrical accessories, it is still usual to include a firmer bevel-edge wood chisel in the tool kit (Fig. 1.9). This has a drop forged alloy steel blade with a wood or plastic handle.

The chisel should be kept in a rack, never loose in the box with other tools, as this would almost certainly cause damage to the cutting edge.

To sharpen, a thin film of mineral oil is spread evenly over the surface of an oilstone and, with the edge resting at an angle of 30°

9

to the stone, the blade is moved backwards and forwards over the surface, taking care not to vary the angle or the pressure. Turning the blade over with the back flat against the stone, the burr is removed by circular movements.

FIG. 1.9. Wood chisel with oil-stone and oil-can.

Cold Chisels

A cold chisel is made from high-quality steel throughout and has a specially hardened and tempered cutting edge. There are many different types (Fig. 1.10), those principally used in electrical work being 203 mm long by 16 mm octagon for cutting walls, etc.

The electrician's bolster chisel, used for taking up floorboards, is 229 mm long by 16 mm octagon, with a 57 mm cut.

Associated with drilling chisels is the universal Rawltool holder. This is made in three sizes: (1) for drills nos. 6–10 (used with a 0·2–0·3 kg hammer); (2) for drills nos. 12–20 (used with a 0·4–0·7 kg hammer); (3) for drills nos. 22 upwards (used with a 1·0–1·5 kg hammer). Drills are easily removed from the holder when required by the use of an ejector pin.

When drilling with a cold chisel or Rawltool, the tool should be tapped with the hammer and turned slightly after each tap (Fig. 1.11). This presents a fresh surface to the cutting edge at each blow, and assists the clearing of debris.

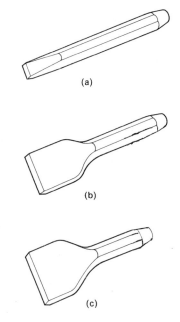

(a)

(b)

(c)

FIG. 1.10. Cold chisels: (a) flat, (b) chaser, (c) bolster.

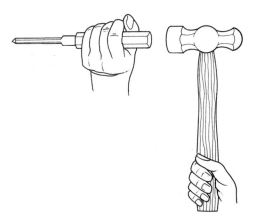

FIG. 1.11. Using Rawlplug drill and holder.

Drill Braces, Bits, and Drills

A brace is used to rotate a bit or drill fitted in the holder, or chuck. Normally, a bit is used for drilling wood and a drill for drilling metals. The common ratchet brace has a U-shaped cylindrical arm and is fitted with a hardwood head and handle. The hand drill is driven by means of an iron bevel wheel, and has a side and top handle in addition to the driving handle. The breast drill is similar to the hand type, except that a breast plate takes the place of the top handle and the speed of rotation can be changed by alternative bevel wheels (Fig. 1.12).

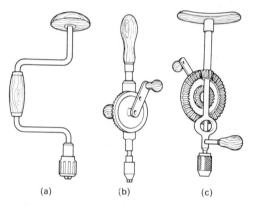

(a)   (b)   (c)

FIG. 1.12. (a) Brace, (b) hand drill, (c) breast drill.

Woodworking braces are available with from 100 mm to 250 mm sweeps. The electrician's brace is usually that with the smallest sweep. The jaws of the chuck are adjustable to take bits and drills with square, round, and tapered shanks from 3 to 12 mm sizes. Two popular types are twist bits and centre bits. Adjustable or "expansion" bits are sometimes used (Fig. 1.13).

Undoubtedly the most commonly used drill is the twist drill. It is supplied in two grades, carbon steel and high speed. If carbon steel drills are operated at high speed the cutting edges are damaged.

The size of a twist drill is marked on the shank. There are alternative methods of specifying drill sizes: (1) lettering (A–Z, representing 6 mm diameter to 10·5 mm diameter); (2) numbering (no. 60 to no. 1, representing 1mm diameter to 5·8 mm diameter); (3) metric (directly in millimetres).

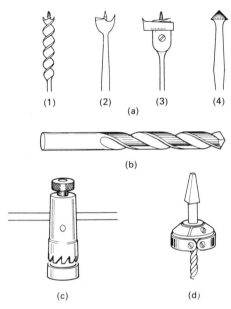

Fig. 1.13. (a) Bits: (1) twist, (2) centre, (3) expansion, (4) countersink. (b) Twist drill. (c) Cooke's hole-cutter. (d) Enox ring saw.

High-speed twist drills can be sharpened on a grindstone or emery wheel. Great care must be taken to ensure that both cutting edges are at the same angle to the axis of the drill and are of equal length, also that the metal is not allowed to overheat. Drills should not be dipped into water to cool them after grinding, as this causes surface cracks. Overheating is indicated by a blue tinge.

As an emergency measure, bits and drills may be sharpened using a small file when a power grindstone is not available.

Apart from twist drills, there are countersink drills for making a conical recess to accommodate the screw-head, counterbore drills for recessing to accommodate a bolt-head, and masonry drills for drilling walls.

13

Drills are best stored in a case or rack containing grooves or holes of appropriate diameters clearly marked with the size.

Auger

Augers are useful for boring relatively large-diameter holes in wood when movement is restricted. They are constructed wholly of steel and described as having "a bright scotch screw and a barrel eye". The barrel eye at the end of the handle enables additional leverage to be obtained after insertion of a rod or conduit (Fig. 1.14).

FIG. 1.14. Auger.

General

An occasional wipe with an oily rag will prevent the formation of rust in iron or steel parts. If rust has started to form, it should be removed with fine emery cloth or soft steel wool before oiling. Wooden parts of tools can be kept in good condition by rubbing linseed oil over the surface.

In addition to the tools already mentioned, the electrician may need for sheathed cable work a trowel, rule, chalk-line, plumb-line, and spirit-level (Fig. 1.15).

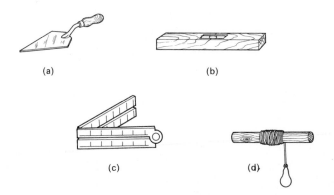

(a)

(b)

(c)

(d)

FIG. 1.15. (a) Trowel, (b) spirit-level, (c) folding-rule, (d) plumb-line.

Questionnaire No. 1

1. In stripping the insulation from tough-rubber- or PVC-sheathed cable, the knife should be held . . . .
2. The hammer should be held . . . .
3. To avoid damaging the screw-head, the screwdriver must have . . . .
4. Side-cutting pliers are used to . . . .
   Gas pliers are used to . . . .
   Diagonal-cutting nippers are used to . . . .
5. When not in use, the teeth of a tenon saw should be protected by . . . .
6. A plaster-saw is used to . . . .
7. To prevent damage to the cutting edge, a wood chisel should be stored . . . .
8. After each blow with a hammer, the cold chisel or Rawltool should be . . . .
9. Operating a carbon steel drill at too high a speed causes . . . .
10. The following kinds of oil are used in connection with tools: for sharpening . . . . oil; for preserving wood parts . . . oil.

TOPIC NO. 2

Safety

If the fuse shatters as soon as you switch "on", you can be reasonably sure that there is something wrong with your circuit.

CERTAIN people are said to be "accident prone", which means that they seem to have more accidents than other people. Yet investigations into the causes of accidents show that the majority could have been avoided if more care had been taken beforehand. In other words if safety had been the first consideration.

Typical "safety-first" hints applying to electrical work are given in this topic, followed by recommendations as to the action to be taken if an accident does occur.

Electrical Safety Precautions

1. Never attempt an electrical repair job with the "juice" on.
2. When repairing fuses always use the correct size of fuse element, never larger.

16

3. If the fuse shatters as soon as you switch "on" you can be sure that there is something wrong with your circuit.
4. Avoid trailing flexible leads whenever possible and beware of those that are essential.
5. Do not neglect protection on temporary work.
6. Remember that it does not pay to play with electricity; a practical joke that leads to injury or death to a fellow student is not funny.

Fire Fighting

When fire breaks out it is necessary to act quickly and coolly. The immediate steps are: summon the fire brigade (set off the fire alarm if there is one), direct people to a safe route out of the building, and start putting out the fire.

Fire requires fuel, a high temperature, and a supply of air. Therefore in fighting fire the object is to remove one or more of these requirements (Figs. 2.1 and 2.2). The removal of fuel is self-

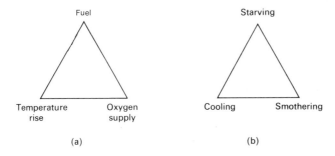

FIG. 2.1. (a) Combustion requirements, (b) extinction requirements.

explanatory—flammable material in the region of the fire should be taken away. Lowering the temperature is achieved by playing water except in the case of oil or electrical fires. An example of restricting the supply of air is the use of a fire blanket.

There are various types of fire-extinguishers, but the manufacturers' recommendations should be carefully followed, particularly when dealing with oil or electrical fires.

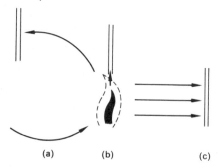

Fig. 2.2. Fire may spread by: (a) convection, (b) conduction, (c) radiation.

First Aid

Ambulance men have said that the majority of fatal accident victims die from shock. Shock, in medical terms, is over-stimulation of the nerves. It can result from a wound or from the passage of electric current through the body. In mild cases it causes pallor, trembling, feeble pulse, respiration and heart action, and a lowering of the body temperature. The treatment for slight shock is to lay the patient down, cover with blankets or coats and, with certain exceptions such as a seriously injured person who might be bleeding internally, provide hot tea or coffee (sweetened, unless the patient is diabetic). Never give alcohol. Obviously if there is considerable bleeding skilled assistance should be summoned quickly to stop bleeding.

First-aiders will know that fractured limbs should be immobilized, that a tourniquet should be applied for periods of up to 15 minutes in cases of severe cuts, and that a sterile dressing should be applied rapidly over burns to exclude the air.

Artificial Respiration

A person may be suffering from an electric shock so severe that breathing has stopped, in which case the first-aid treatments just described would not be used. Artificial respiration must be applied (Fig. 2.3). It must be applied immediately, otherwise the victim's chances of surviving will soon disappear.

The recommended procedure is:
1. Switch off the electricity supply when practicable and, if necessary, release the person from contact with live portion (do not touch with bare hands while connected, but drag clear using clothing or other insulating material).
2. Start artificial respiration as follows, and send for doctor and ambulance.
3. Place the victim on his back and place a folded coat or other clothing under the shoulders so that the head falls well back.
4. Pull the chin forward, making sure the air passage is clear, pinch the nostrils, breathe steadily through the mouth into the victim's chest until his chest rises, then move clear to allow exhalation.
5. Repeat at the rate of 12–15 times a minute until the victim revives or the doctor gives fresh instructions.

FIG. 2.3. Artificial respiration (acknowledgements to the Order of St. John).

Regulations

There are regulations designed to ensure safety to people and property. The Institution of Electrical Engineers Regulations cover the electrical equipment of buildings in general. The Electricity (Factory Act) Special Regulations, 1908 and 1944, apply to factories using electricity. Electricity Supply Regulations, 1937, include a certain amount that applies to consumers' installations. Codes of Practice, issued by the British Standards Institution, deal with installations and equipment of various kinds.

Strict observance of these regulations is essential to safety.

General Safety Precautions

1. Use things only for the purpose for which they are intended.
2. Place tools securely and safely (for instance, where they will not fall or where people will not step on or fall over them).
3. Do not wear loose clothing or long hair.
4. Keep cutting edges sharp; many wounds result from trying to force blunt tools to cut.
5. Make certain that step-ladders are firm, suitable, and correctly positioned for the job in hand.

Questionnaire No. 2

1. Before attempting an electrical repair you should be sure ....
2. A sharp tool is often safer to use than a blunt one because ....
3. Electrical equipment of buildings is covered by the regulations of ....
4. Shock can result from either of the following two causes .... or ....
5. An injured person should never be given ....
6. When a person has stopped breathing it is essential to apply immediately ....
7. To release a person from contact with a live portion ....
8. If fire breaks out the immediate steps are ....
9. Fires are extinguished by removing one or more of the following: ...,
10. Oil or electrical fires must not be treated with ....

Cables

The purpose of the insulation is to prevent leakage of electricity from the conductor.

A CABLE consists of at least two parts: *conductor* and *insulation* Some cables have, in addition to these two parts, some form of protection against mechanical damage, such as a sheath of PVC or copper or armouring.

Copper has been commonly used for conductors mainly because it is a good conductor of electricity, which means that it offers little opposition, or *resistance*, to an electric current. To a lesser extent aluminium, which has a higher resistance than copper, is employed as cable conductor.

Materials used for cable insulation include: rubber, polyvinyl chloride (PVC), polythene, mineral magnesium oxide, and paper tape impregnated with oil. The purpose of the insulation is to prevent leakage of electricity from the conductor, in other words it has the opposite function to that of the conductor. The higher the voltage applied to the conductor, the thicker must be its insulation. Cables for ordinary wiring have voltage ratings of insulation of 600/1000 or of 600 or 1000 for mineral-insulated cables.

Cable coverings, sheaths, and armouring, when included, prevent damage to insulation and consequently to conductors both when installing and in service. For coverings, waxed cotton braid and impregnated cotton tape are sometimes used; for sheaths, tough rubber, polythene, PVC, copper, aluminium, and lead; for armouring, galvanized steel wire or tape.

Cables commonly used in electric wiring include the following:

VRI (vulcanized-rubber insulated)

This consists of a copper conductor, rubber insulation, and a protective covering of cotton braid. The rubber is vulcanized to make it mechanically stronger. Because the sulphur, which is added to the rubber in the vulcanizing process, attacks copper, the conductor is "tinned" (coated with a thin layer of tin) (Fig. 3.1).

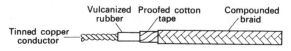

FIG. 3.1. VRI taped and braided cable.

PVC (polyvinyl chloride) insulated

This also has a copper conductor but, instead of vulcanized rubber, polyvinyl chloride is used for the insulation (Fig. 3.2). This type of insulating material is tougher than rubber, therefore mechanical protection in the form of braid is not required. Also, as there is no sulphur in the material, it is unnecessary to tin the conductor. Polyvinyl chloride is not as good an insulator as rubber and has the disadvantage that it hardens in cold temperatures. On the other hand, it is unaffected by certain liquids which attack rubber, and this form of insulation is now listed in the *IEE Tables of Current Rating* to the exclusion of rubber.

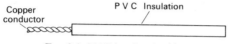

Fig. 3.2. PVC-insulated cable.

TRS (tough-rubber-sheathed)

This is obtained with one, two, or three cores contained in a sheath of toughened rubber. Each core has a tinned copper conductor with rubber insulation (Fig. 3.3). An uninsulated copper earth-continuity conductor is included within the sheath.

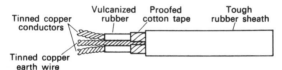

Fig. 3.3. TRS two-core cable.

PVC (polyvinyl chloride) insulated and sheathed

This is normally similar to tough-rubber-sheathed except that PVC is used instead of rubber for insulation and sheath (Fig. 3.4). The sheath is made of a tougher grade of PVC than that used for insulation. Steel wire armouring is applied to some types with an overall PVC sheath. Butyl rubber, e.p. rubber or silicone rubber is sometimes used for insulation in place of PVC. In the larger sizes

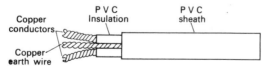

Fig. 3.4. PVC-sheathed cable.

23

PVC-sheathed aluminium conductors are sometimes employed. As in the case of TRS, an earth-continuity conductor is included.

MICS (mineral-insulated copper-sheathed)

This type of cable, made by Pyrotenax Ltd., British Insulated Callenders Cables Ltd., and others, has an unstranded copper conductor with insulation consisting of mineral magnesium oxide powder, which can withstand high temperatures, and a tubular sheath of copper (Fig. 3.5). It can be obtained with one, two,

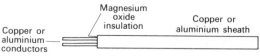

FIG. 3.5. Mineral-insulated cable.

three, four, and seven cores and, if required, an overall covering of PVC to prevent possibility of action when in contact with certain chemicals (e.g. when buried or exposed to corrosive atmosphere). Seals and glands are used in conjunction with MICS cables at all terminations because the mineral insulation absorbs water.

MIAS (mineral-insulated, aluminium-sheathed)

This resembles MICS cable, except that conductor and sheath are of aluminium instead of copper. Because aluminium is not as good a conductor of electric current as copper MIAS cables are of larger diameters than those of MICS cables designed to carry the same current. Therefore, the corresponding seals and glands are larger than those used with MICS. Aluminium corrodes in damp situations when in contact with certain materials, especially copper.

PILSA (paper-insulated, lead-sheathed, armoured)

This is used a great deal by supply authorities and occasionally by electrical contractors, especially for underground runs between buildings. It consists of copper conductors insulated with lappings of oil-impregnated paper tape, sheathed with lead, and armoured with steel wire or tape. Bitumen-impregnated jute is applied between lead sheath and armouring and over the armouring (Fig. 3.6).

The cable must be effectively sealed at all terminations, which also incorporate a gland to fit over the sheath and an armour clamp to grip the steel wires or tapes.

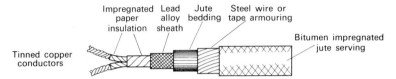

FIG. 3.6. Paper-insulated, lead-sheathed, armoured cable.

Questionnaire No. 3

1. The two essential parts of a cable are . . . and . . . .
2. A material is said to be a good conductor of electricity because its resistance is . . . .
3. Cable insulation is used to . . . .
4. The difference in voltage ratings of insulation is in the . . . .
5. Armouring of cables consists of either . . . or . . . .
6. Copper conductors of VRI cables are tinned because . . . .
7. Two disadvantages of PVC in comparison with rubber are . . . .
8. Mineral magnesium oxide insulation has the advantage that . . . .
9. Seals and glands used with MIAS cable are larger than those used with corresponding MICS cables because . . . .
10. In damp situations the aluminium sheath of cables must be protected against corrosion especially when there is likelihood of contact with . . . .

25

TOPIC NO. 4

Conductor Terminations

There should never be any mechanical strain on a termination.

THE *IEE Regulations* require that terminations of cable conductors shall be accessible for inspection and mechanically and electrically sound. In other words, they should not be difficult to get at, they should not be such that they could be easily dislodged, and they should provide good electrical connection.

When terminating cables at accessories such as ceiling roses, lampholders, switches, joint boxes, etc., it is very important that there should be effective contact between conductors and terminals, and that all of the wires forming a conductor are anchored. If a connection is loose, or some of the strands are not secured, over-heating will occur, and this is a possible source of danger. Efficient connection depends on tight contact over as large a surface-area as possible.

There should never be mechanical stress on a termination. Due to the fact that a conductor becomes heated and expands each time current flows through it, any pull exerted is liable to loosen the connection. In cutting off any surplus, care should be taken to leave a sufficient length of conductor for the termination while avoiding too much slack.

Stripping Cables

In preparing cable ends for entry into terminals, it is necessary to remove any mechanical protection and insulation in order to bare a sufficient length of conductor for the termination. This stripping process has already been described. It is also necessary to remove the braid and tape of VRI cable for an adequate distance from the end of the insulation. The reason for this last operation is that tape and braid are included in the cable for mechanical protection and not for additional insulation. Under damp conditions it is possible for these materials to conduct, so that if they are in contact with the conductor they may cause leakage. PVC insulated cable has no braid or tape; therefore in this case it is necessary merely to bare the required length of conductor.

As TRS cable consists of rubber-insulated cores contained in a sheath of toughened rubber, the first step in stripping is to remove an adequate amount of sheath, allowing at least 12 mm in addition to the required length of bare conductor. A similar procedure is adopted in

27

stripping PVC-sheathed cable. Precautions against corrosion are necessary when a termination in a damp situation involves contact between aluminium and copper or brass.

In stripping mineral-insulated cables special tools are used to remove the metal sheath and seals and glands are fitted. These terminations are dealt with later as a separate topic, as are those involved with paper-insulated and PVC-insulated and armoured cables. When a cable termination is to take the form of a cable socket, or lug, soldering or crimping is required and these processes also are treated individually in due course.

Great care is necessary when stripping flexible cords as the conductors are made up of very fine strands which are easily cut or broken off with rough handling.

If a length of flex. cord is extended, the connection must be by means of a non-reversible coupler shrouded in either earthed metal or incombustible insulating material, so that no live parts can be exposed in service.

When using a knife for stripping, the blade should be held at an acute angle to the surface of the insulation (Fig. 4.1). This is because,

FIG. 4.1. Knife should be at an acute angle when stripping cable.

in the event of an excessive cut, it is less harmful to take a thin slice out of the conductor than to make a deep cut (Fig. 4.2). A nicked

(a)   (b)

FIG. 4.2. It is less harmful to take a thin slice out of a conductor, (a), than to make a deep cut, (b).

conductor will break after being bent a few times. Moreover, a cut will reduce the effective cross-sectional area of the conductor to a greater extent than a corresponding slice.

Before using a stripping tool it is important to make sure that it is of a well-designed type and that it is correctly adjusted for the size of conductor.

In terminating a stranded conductor the strands should be twisted together after removal of the insulation, taking great care not to damage the wires in the process.

Damp Situations

If cables are terminated in damp situations, precautions must be taken against the possibilities of corrosion and electrolytic action of metal sheaths, conduits, etc. To comply with the *IEE Regulations* it is necessary in all cases for the cores of sheathed cables from which the sheath has been removed and for non-sheathed cables at the termination of conduit, duct, or trunking to be enclosed in incombustible material and in a damp situation the enclosure must be damp-proof and corrosion-resistent. Thus cable ends which terminate at or are looped into an accessory or lighting fitting in such a situation may have to be in a sealed box.

Special care is always necessary when terminating cables and flexible cords in situations liable to dampness, as leakage is more serious under such conditions.

High Temperatures

It should be ensured that cables do not terminate where they are likely to be exposed to excessive temperatures, for example within an enclosed light fitting or heating appliance. Particular caution is necessary in the case of polythene-insulated cables, as this melts very readily. Maximum operating temperatures to which the insulation and sheath of cables and flexible cords should normally be subjected are:

Rubber compound, or PCP	60°C
PVC compound	70°C
Mineral (with 80°C terminations) varnished cambric or impregnated paper	80°C
Butyl or e.p. rubber	85°C
Silicone rubber	150°C

The two last-named types of flexible cords are recommended for the connections between ceiling roses and lampholders in pendant lighting fittings in situations where tungsten filament lamps are to be used. Inside a lighting fitting or appliance the cable or flexible cord should have insulating sleeves or heads suitable for the temperatures likely to be encountered and fitted over individual cores.

Terminals

Probably the most frequent criticism by electricians of accessories is the size of the terminals or the ease or otherwise with which the end of the conductor can be secured. Terminals used in electrical installation work range from the pillar type to screw-heads and nuts (Fig. 4.3). Thus the preparation of the conductor end will depend on the kind of terminal to which it is to be connected.

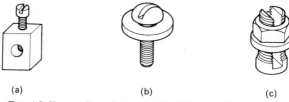

(a)   (b)   (c)

FIG. 4.3. Types of terminals: (a) pillar, (b) screw-head, (c) nut.

Before entering a single conductor into a pillar terminal, it is preferable, providing there is room, to double the conductor end back on itself, thereby providing a larger surface of contact. When two or more conductors have to be secured to the same terminal they should be twisted together before entry (Fig. 4.4). This ensures

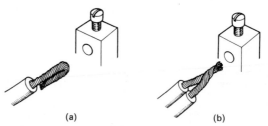

(a)   (b)

FIG. 4.4. Entry into pillar terminal of (a) single conductor, (b) two conductors.

that contact between the conductors does not depend on the terminal screw. A similar procedure should be followed, if possible, when terminating conductors in split screws.

Conductor ends which have to be looped under screw-heads or nuts are formed into an eye using round-nosed pliers. The eye should be of slightly larger diameter than the screw shank but smaller than that of the screw-head, nut, or washer. It is usual to arrange the eye over the screw in such a way that rotation of the screw-head or nut tends to tighten the loop of the eye and not to unwind it (Fig. 4.5).

FIG. 4.5. Conductor end looped under screw terminal.

Questionnaire No. 4

1. Terminations of cable conductors must be . . . .
2. A loose connection will cause . . . .
3. The reason that the braid and tape of VRI cable-ends is removed from the end of the insulation is . . . .
4. Great care is necessary when stripping flexible cords due to the fact that . . . .
5. When using a knife for stripping, the blade should be held at an . . . angle to the surface of the insulation because . . . .
6. In terminating a stranded conductor, the strands should be . . . .
7. Special care is necessary when terminating cables and flexible cords in a damp situation as . . . .
8. The maximum operating temperatures to which rubber or PVC compound insulation should be subjected are . . . for rubber, and . . . for PVC.
9. Twisting conductors together before entering them in the same terminal ensures . . . .
10. When looping conductors under screw-heads or nuts, it is usual to arrange the eye over the screw in such a way that . . . .

Insulating Sheathed Wiring

Tough-rubber-sheathed cables should never be exposed to direct sunlight.

AN ELECTRIC wiring system consists of those parts which form electric circuits. That is, the conductor, insulation, mechanical protection, and accessories used for fixing, joining, and terminating.

Systems

Systems can be roughly grouped into those which are (a) single-core cables, such as PVC-insulated, protected by enclosure in conduit or trunking; (b) single-, twin-, and multi-core metal-sheathed cables, such as mineral-insulated copper or aluminium-sheathed; and (c) twin- and three-core insulating sheathed cables, such as

32

rubber- or PVC-insulated and sheathed, installed without protection except in places where they are particularly liable to mechanical damage (e.g. through floors or walls).

This topic concerns the use of twin- and three-core rubber or PVC- insulated and sheathed cables (Fig. 5.1).

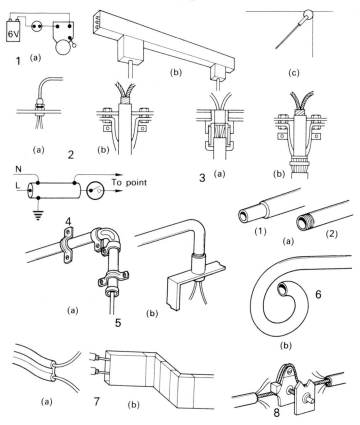

FIG. 5.1. Examples of wiring systems: (1) bare or lightly-insulated conductors: (a) extra-low voltage, (b) bus-bar trunking, (c) collector-wire. (2) Metal-sheathed cables: (a) mineral-insulated, (b) paper-insulated lead-sheathed. (3) Armoured: (a) PVC-sheathed, (b) paper-insulated lead-sheathed. (4) Earthed concentric. (5) Metal conduits: (a) plain, (b) screwed. (6) Non-metallic conduits: (a) rigid, (b) flexible. (7) Prefabricated: (a) harness, (b) skirting. (8) Catenary.

Advantages of insulating sheathed wiring as compared with other wiring systems are that it is comparatively cheap to install and can often be easily concealed or run unobtrusively on surfaces.

Methods

The methods of wiring lighting points in this system entail the use of either joint boxes or three-plate ceiling roses (Fig. 5.2). In the

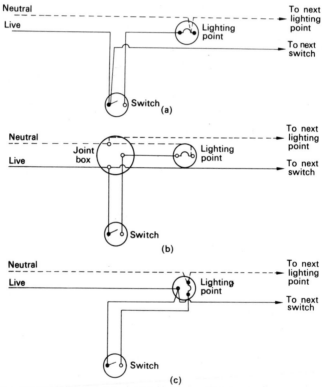

FIG. 5.2. Methods of wiring: (a) looping-in, (b) joint box, (c) three-plate ceiling rose (earthing omitted for simplicity).

joint-box method live and neutral conductors are taken from the distribution point to a terminal in a joint box fixed in a convenient position, and thence to lighting points and switches, etc., joints in switch wires being made in other terminals within the joint box. In the three-plate ceiling-rose method one terminal in the ceiling rose is connected to the live conductor, and the other two terminals (from which the flexible cord is taken) to the switch wire and neutral. The *IEE Regulations* require that the live terminal is screened, otherwise there would be a danger that the householder may assume that all terminals are dead when he switches off. An earthing terminal must be provided.

Before starting any wiring, the route and arrangement should be carefully planned. It is best to group as many cables together as possible and to keep the runs straight. Tough-rubber, PVC- and other non-metal sheathed cables should never be exposed to direct sunlight.

Fixing Sheathed Wiring

Tough-rubber-sheathed and PVC-sheathed cables are usually fixed by buckle clips and saddles (Fig. 5.3). It is good practice to

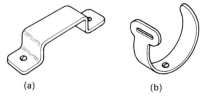

(a)   (b)

Fig. 5.3. Methods of fixing TRS and PVC-sheathed cables: (a) saddle, (b) buckle clip.

secure single runs by clips and multiple runs by saddles. For fixing clips and saddles to most wall surfaces, brass screws and plugs are used, but it is permissible to use brass pins for fixing direct to woodwork.

To secure cable by buckle clips, the run is marked out and the clips fixed at appropriate intervals. The maximum spacing in accessible positions for rubber- or PVC-sheathed cables varies from 250 mm horizontally and 400 mm vertically for the smaller sizes (not exceeding 9 mm overall diameter) to 400 mm horizontally and 550 mm vertically for the larger sizes (up to 40 mm overall diameter). The cable is laid across the fixing screw or pin and the ends of the clip are lifted, the tail is pushed through the eyepiece, drawn up tight, then bent back on itself.

When securing by saddles, one screw or pin can be loosely fixed in all saddles in the line of the run, so that the cable or cables can be tightened and slipped into each saddle in turn, and the fixing completed. In the case of multiple runs using metal saddles, it is possible to obtain strip or tape which can be cut up, bent, and drilled to suit the total width of cables. Plastic cable clips incorporating specially-hardened fixing pins are also available.

Cable Manipulation

It is much better to run cables from a revolving reel (Fig. 5.4) than direct from the coil laying on the floor. This avoids kinking. Whenever possible, the cables should be run *down* walls rather than up.

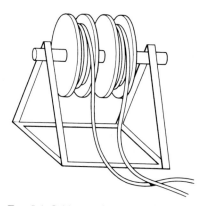

FIG. 5.4. Cables run from revolving reels.

If cables have to be run on recently papered or finished surfaces, it is useful to fit paper strip under the cables as they are fixed, so that the painting of cables can be carried out without risk of marking the wall, the paper strip being torn away after the paint has dried.

At bends, the minimum internal radius for cables, between 10 mm and 25 mm overall diameter, is four times the overall diameter of the cable. In turning a right angle with tough-rubber-sheathed cable, it may be found easier to turn the cables completely over at corners. Tough rubber must not be exposed to direct sunlight; cables run in a situation where this is possible should be provided with a special protective covering.

Joint boxes should be fixed in accessible positions. All terminations should take place in incombustible enclosures, the sheath being taken right into them.

When there is any risk of mechanical damage, the cables should be protected by bushed conduit or capping, and when cables pass through structural metalwork the hole should be suitably bushed to prevent abrasion.

Questionnaire No. 5

1. Advantages of insulating sheathed wiring as compared with other wiring systems are . . . .
2. In the joint-box method of wiring, live and neutral conductors are taken . . . .
3. In the three-plate ceiling-rose method of wiring, one terminal in the ceiling rose is connected . . . and the other two terminals to the . . . and . . . .
4. When planning the route and arrangement of wiring it is advisable to . . . .
5. When fixing TRS and PVC-sheathed cables, it is good practice to secure . . . by clips, and . . . by saddles.
6. To secure cable by buckle clips, the cable is laid across the fixing screw or pin and . . . .
7. To secure cables by saddles, one screw or pin can be loosely fixed in all saddles in the time of the run, so that . . . .
8. Running cables from a revolving reel rather than direct from the coil laying on the floor avoids . . . .
9. The minimum internal radius of a bend in TRS or PVC-sheathed cable is . . . .
10. When there is risk of mechanical damage, cables should be . . . .

Flexible Cables and Cords

A flexible cable is a cable designed to afford flexibility.

A FLEXIBLE CABLE, as the name implies, is a cable specially designed to afford flexibility. The cable conductor therefore consists of a number of fine wires. If the cross-sectional area of each conductor does not exceed 4 mm², it is known as a *flexible cord*. A 4 mm² conductor consists of 56 wires, each of 0·3 mm diameter, and is rated at 25 A; in other words it is, by domestic standards, a relatively large size. Thus, most of the flexibles with which we are likely to be concerned in small installations will be flexible cords.

Most types of flexible cord are available with the alternatives of two, three, and sometimes four cores, with either medium or heavy insulation. In some, the cores are twisted; in some, side by side. They may be provided with flat sheaths or circular sheaths.

Some of the commonest types are illustrated in Fig. 6.1 and described as follows:

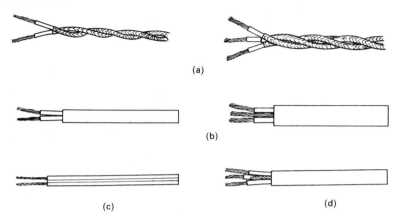

(a)

(b)

(c)   (d)

FIG. 6.1. Common flexible cords: (a) twisted, (b) flat, (c) figure-8 or double D, (d) circular.

VRI Braided Flexibles

These have tinned copper strands insulated with vulcanized rubber. In the "twin-twisted" types with medium insulation, the cores are braided with glacé cotton and twisted together. The standard cotton braiding is maroon in colour. Special types of flexible cord with artificial silk braiding in alternative colours are also manufactured.

VRI braided "circular" cords have the cores twisted together and cotton padding included to make up a rounded section with glass cotton braiding overall in either maroon, black, or white. They can be obtained with two or three cores, medium or heavy insulation.

TRS Flexibles

In addition to tinned-copper-stranded conductors insulated with vulcanized rubber, this type has a circular sheath of toughened rubber generally either black or white in colour. It is commonly used in situations liable to dampness, such as kitchens and bathrooms. Typical alternatives are two, three, or four cores, medium or heavy insulation.

PVC-insulated Flexibles

Typical flexible cords under this heading are: single, twin-twisted, and figure-eight. Some have plain stranded conductors; others tinned copper. The 300/300 V grade insulation is available in a wide range of colours.

PVC-insulated and Sheathed Flexibles

The stranded conductors of these, generally untinned, are PVC-insulated with the cores twisted together, padded with cotton and provided with a circular PVC sheath. This type is available with two, three, or four cores. Standard sheath colours are grey, black, white, and brown.

Special Flexible Cords

Non-kinking domestic flexibles have cores comprising fine copper wires insulated with vulcanized rubber. The cores are twisted together with strengthening cords in the interstices. The whole is enclosed in a circular rubber sheath and glacé cotton braided overall (Fig. 6.2).

For electric cookers or other appliances requiring a flexible which is resistant to heat and impervious to moisture, glass-fibre

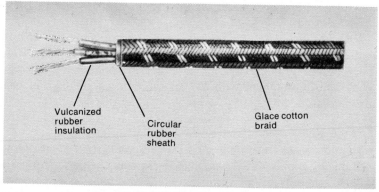

Vulcanized
rubber
insulation

Circular
rubber
sheath

Glace cotton
braid

FIG. 6.2. Non-kinking domestic flexible.

insulated copper may be used. This has a tinned-copper conductor lapped with impregnated glass fibre and covered with glass-fibre braid (Fig. 6.3).

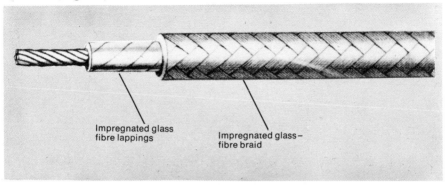

Impregnated glass fibre lappings

Impregnated glass– fibre braid

FIG. 6.3. Heat-resisting flexible.

Other heat-resisting cords available are those insulated with butyl- and silicone-rubber.

Questionnaire No. 6

1. A flexible cable is a cable . . . .
2. A flexible cord is a flexible cable with a conductor having a cross-sectional area not exceeding . . . .
3. The largest size of flexible cord has a conductor consisting of . . . wires, each . . . in diameter and having a rating of . . . .
4. VRI braided flexible cords consist of . . . .
5. VRI braided "circular" cords have the cores . . . .
6. TRS flexible cords consist of . . . .
7. TRS flexible cords are commonly used in . . . .
8. Three typical types of PVC-insulated flexible cords are . . . .
9. PVC-insulated and sheathed flexible cords consist of . . . .
10. Two special types of flexible cord are: (1) non-kinking domestic, which consists of . . ., and (2) heat-resistant, which consists of . . . .

Conductor Sizes

The function of a conductor is to carry an electric current.

WE HAVE seen that the function of a conductor is to carry an electric current. The ease with which the conductor conveys a current depends on its cross-sectional area which, in the case of a round conductor, is equal to 0·7854 multiplied by the diameter squared, i.e. $[(\pi/4)d^2]$. The smallest size of cable normally used in electrical installation work has a conductor diameter of 1·13 mm. Its cross-sectional area, therefore, is

$$0.7854 \times 1.13 \times 1.13,$$

which equals approximately 1 mm^2.

In theory, the current-carrying capacity of a conductor is inversely proportional to the cross-sectional area. This would mean that, for example, a 16 mm^2 conductor could carry 4 times the current that a 4 mm^2 conductor could carry. However, when dealing with the majority of insulated cables the maximum permissible current they carry is based largely on the temperature rise of the insulation, which depends on a number of factors.

Stranded Conductors

You will have noticed that some cables such as VRI and PVC-insulated, have conductors which are made up of several wires twisted or laid up together. The direction in which the wires, or strands, are twisted is called the *lay* (Fig. 7.1).

(a)   (b)

FIG. 7.1. Lay of stranded conductor: (a) right-hand, (b) left-hand.

Let us explain why cable manufacturers go to the trouble and expense of forming the conductor of a number of strands. If we first of all consider a conductor which consists of a single wire, it is obvious that the greater the cross-sectional area of the wire, the more difficult it is to bend. In fact, this still applies when more than one strand is used to make up the conductor. For example, it would be found easier to bend sixteen wires each of 0·5 mm^2 area than to bend one wire of 8 mm^2 area. In other words, forming a conductor of strands makes it more flexible.

Flexible cables and cords, as we have seen, have conductors that consist of a large number of very fine wires. Ordinary wiring cables are not required to be as flexible as these but when they are to be drawn into conduits, trunking, or ducts, they must at least

43

be capable of bending easily without being damaged. Therefore the conductors of most VRI and PVC-insulated cables, which are designed for drawing in or enclosing within an insulated sheath, are stranded.

In electrical installation work, it is usual to express the sizes of stranded copper and aluminium conductors by giving the number of strands followed by the diameter of one strand. The two quantities are separated from one another by an oblique stroke. Thus, 7/0·85 indicates a conductor having seven strands each having a diameter of 0·85 mm. Similarly, a 7/2·14 conductor consists of seven strands each of which has a diameter of 2·14 mm.

Engineers concerned with large-stranded cable conductors (e.g. service cables) customarily express size by cross-sectional area. Also, sizes of cables necessary for a particular purpose are calculated in this way. The *IEE Tables of Current Ratings* cater for this by listing stranded cable sizes up to 19/1·53 mm in cross-sectional area as well as in number and diameter of strands.

Use of Wire Gauge

The diameter of a strand can be found by testing with a wire gauge (Fig. 7.2.), which gives the size in standard wire gauge (s.w.g.), and then using the conversion table below:

s.w.g. no.	Nearest strand diameter (mm)
12	2·61
13	2·36
14	2·11
15	1·83
16	1·63
17	1·32
18	1·12
20	0·91
22	0·74

It can be seen from the manufacturers' cable lists that the number of strands which go to make up a cable conductor is one of the following: 7, 19, 37, 61, 91, 127. From Fig. 7.3, showing the various

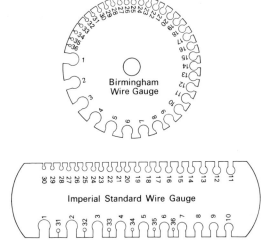

FIG. 7.2. Wire gauges.

FIG. 7.3. Arrangement of strands in a cable conductor.

arrangements, you will understand why these numbers are chosen.

Questionnaire No. 7

1. The cross-sectional area of a conductor with a diameter of 1·78 mm is. . . .
2. A 4 mm² conductor could carry . . . the current that a 2 mm² conductor could carry.
3. In most cases the maximum permissible current cables carry is based largely on . . . .
4. The direction in which strands are twisted is called the . . . .
5. Forming a conductor of strands makes it . . . .
6. Conductors designed for . . . are stranded.

7. 19/1·53 indicates a conductor having . . . .
8. Cross-sectional areas of stranded cables are given in the *IEE Tables* because . . . .
9. The difference in diameter between one s.w.g. no. and the next is approximately . . . .
10. The reason for the number of strands of a cable conductor being 7, 19, 37, etc., is that . . . .

Consumer's Control Equipment

Usually entailed stumbling across the cellar with a candle to repair a fuse.

ELECTRIC wiring starts at the point where the Electricity Supply Board's service cable terminates in the building, often called the intake position. It is normally here where the consumer's main switch and protective gear is located.

Former Equipment

Years ago it was accepted that for domestic installations the Electricity Board's sealing chamber and cut-out (termed the "service head"), was mounted on a wooden board together with the meters, main switch and fuses, and distribution fuseboard. Often a

47

rudimentary form of main switch was used consisting of two tumbler-type switches with their knobs linked together by a wooden bar. The fuses were either "bow" or open-type cut-outs, or units housed in a wooden case with a glass front. The wiring straggled between the various items of equipment (Fig. 8.1). Add to this the fact that

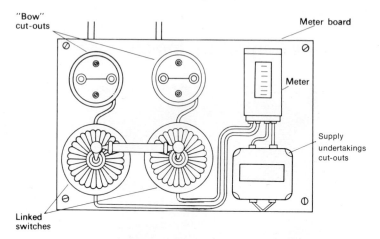

FIG. 8.1. Typical domestic intake arrangement of about 1900.

the equipment was more often than not positioned in the corner of a dark, damp cellar, and you have a reasonably accurate picture of an intake of that period.

The disadvantages of that arrangement are fairly obvious. There were a number of loops of unprotected cable, particularly in view of the fact that there were separate meters for lighting and other purposes. It was not readily accessible and usually entailed stumbling across the cellar with a candle to repair a fuse, switch on or off, or read the meters. Metal parts quickly became rusted or corroded. Additions to the installation generally involved supplementing some of the intake equipment.

Modern Units

In most modern domestic installations, the main control and protective equipment may be combined in one unit. Sometimes the service head and meter is also included. These consumer's units are used in small single-phase installations.

Although there are several alternative forms of consumer's unit, the basic components are a switch and a number of fuseways (or circuit breakers) having the desired ratings, with appropriate provision for neutral connections (Fig. 8.2). Some are metalclad; others all-insulated.

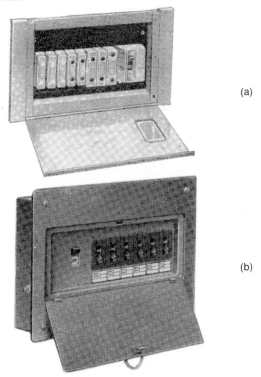

(a)

(b)

FIG. 8.2. Modern domestic control units: (a) with fuses, (b) with miniature circuit-breakers.

One type, which is fairly representative, is obtainable with switch ratings of 30, 45, 60, and 80 A and a maximum number of eight fuseways having alternative ratings of 5, 15, 20, 30, or 45 A. For example, a five-way unit with an 80 A main switch might have one 45 A fuse for a cooker, one 30 A fuse for a ring circuit, one 20 A fuse for a water-heater, and two 5 A fuses for lighting. It is permissible to have the controlling switch of a lower rating than the total fuse rating because it is unlikely that the whole of this load would be switched on at once.

In some cases provision is made in the unit for possible future extensions to the installation. A small additional unit having the desired number of extra ways is connected to the original unit (Fig. 8.3).

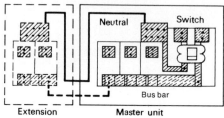

FIG. 8.3. "Extenso"-type distribution unit.

Sequence

The basic requirement is that there should be an adequate means of isolation. Excess current protection in the way of a main fuse can be omitted in the case of the unit provided that the current rating of the cables between the Electricity Board's fuse and the consumer's fuse is not less than that of the Board's fuse, and that the circuit excess-current protection is within the same enclosure as, or fixed immediately adjacent to, the switchgear.

The sequence of equipment at the intake should usually be as follows (Fig. 8.4):

(1) Service fuse and neutral link.
(2) Meter.
(3) Consumer's linked switch or circuit-breaker.
(4) Consumer's fuses.

Fuses

A fuse is defined as "a device for opening a circuit by means of a conductor designed to melt when excessive current flows". It

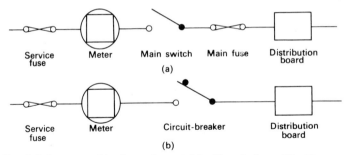

FIG. 8.4. Sequence of intake equipment: (a) with main fuses, (b) with overload circuit-breaker.

normally comprises a fuse link and fuse base. The current rating
of the fuse is the current it will carry continuously without
deterioration (NOT the current at which it will melt).

The term "fuse" covers the whole device. The parts which together make up a fuse are: the fuse element, which is the part which
is designed to melt; the fuse link or carrier, to which the fuse element
is attached and connected at the ends to contacts, and which, when
fitted in a unit, is usually plugged into a base (Fig. 8.5).

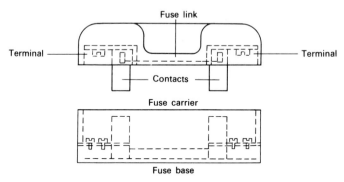

FIG. 8.5. Component parts of a fuse.

Two types of fuses are in common use today: (1) the rewirable
type, in which the fuse element consists of a piece of wire of a
certain gauge and material and which, after blowing and sub-
sequent rectification of the fault, is replaced by a similar piece of

wire, and (2) the cartridge type, often called the high breaking capacity (h.b.c.) type, in which the fuse element is a totally enclosed manufactured component arranged to be plugged into the fuse link (Fig. 8.6). The ratings of cartridge fuses for distribution fuse boards or consumer units are distinguished by colour as follows:

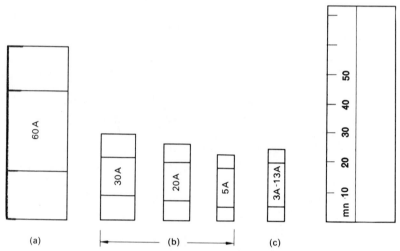

(a) (b) (c)

FIG. 8.6. Sizes of h.b.c. cartridge fuse links: (a) for supply undertakings cut-outs, (b) for domestic-type distribution boards, (c) for fused plugs.

    5 A . . . . white
15 A . . . . blue
20 A . . . . yellow
30 A . . . . red-brown
45 A . . . . green

Advantages of h.b.c. cartridge fuses as compared with rewirable fuses are that as they are not exposed to the air they do not deteriorate, they melt at smaller overloads, and, as they are completely enclosed in fireproof material, there is no risk of fire when they blow. The disadvantage is that they are much more expensive.

Miniature Circuit-breaker Units

A more recent development is the distribution board or consumer's unit using miniature circuit-breakers in place of fuses (Fig. 8.7). The circuit-breaker principle can be considered to be that of an automatic switch operated by an electromagnet. When excess current flows, the current through the coil of the electromagnet is sufficient to cause contacts to separate and thus disconnect the faulty or overloaded circuit.

FIG. 8.7. Section through miniature circuit-breaker.

Advantages claimed for circuit-breakers over fuses are that they are set to disconnect at a definite value of current and there is no risk, as with fuses, of their being replaced by one of a wrong rating, that in the case of an overload the supply can be quickly restored, and that there is immediate indication of operation.

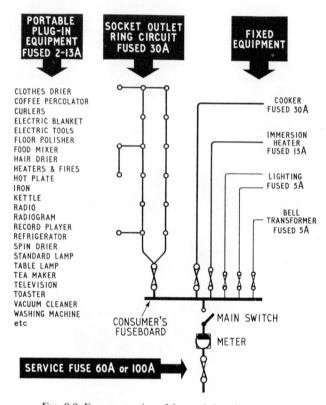

FIG. 8.8. Fuse protection of domestic installation.

Wiring Access

In the most satisfactory designs of fuse board or consumer unit, there is plenty of wiring space provided, and all terminal screws can be easily replaced from the front. The actual arrangement varies with different makes. It is obviously advisable for the neutrals to be connected in the same order as the live conductors, and for an accurate and clearly written circuit list (Fig. 8.8) to be provided.

Questionnaire No. 8

1. Disadvantages of the old intake arrangements were ....
2. Basic components of a consumer's unit are ....
3. It is permissible to have a controlling switch of a lower rating than the total fuse rating because ....
4. A main fuse can be omitted provided that ....
5. The sequence of equipment at the intake should usually be: (1)...; (2)...; (3)...; (4) ....
6. A fuse is defined as ....
7. The parts which together make up a fuse are: (1)...; (2)...; (3) ....
8. Advantages of h.b.c. cartridge fuses as compared with rewirable are: (1)...; (2)...; (3)...; (4) ....
9. Advantages claimed for circuit-breakers over fuses are: (1)...; (2)...; (3) ....
10. In the most satisfactory design of fuse board or consumer unit: (1)...; and (2) ....

Socket Outlets and Plugs

The attendant risks of trailing flexes.

ELECTRIC points intended to supply appliances such as portable fires, vacuum cleaners, table lamps, and television receivers, terminate in socket outlets, although they are often referred to as plug points and sometimes, quite wrongly, as "power" points.

A socket outlet and plug may be considered as two portions of a device for connecting an appliance to the supply. The socket outlet is a fixed part in which the wiring terminates. The plug, which carries metal contacts to connect with corresponding metal contacts in the socket, is normally connected to the appliance by flexible cord or cable.

Radial Circuits

Installations wired on the older system, sometimes called radial circuits (Fig. 9.1) may use round-section plugs and sockets of 2, 5, and 15 A capacity. The 2 A sockets, generally connected to lighting

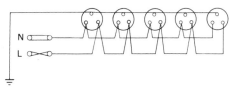

FIG. 9.1. Socket outlets on radial circuit.

subcircuits, are suitable for table and standard lamps. Although rated at 2 A, they must not be used for equipment of more than $\frac{1}{2}$ A loading. Five-ampere sockets may be used for the connection of equipment rated at anything up to 5 A, and similarly, 15 A sockets for equipment rated up to 15 A (Fig. 9.2).

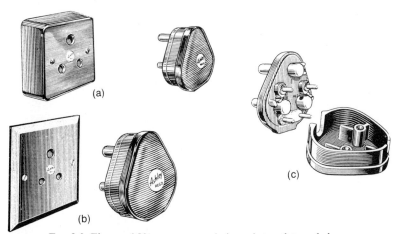

FIG. 9.2. Five- and fifteen-amp round-pin socket outlets and plugs:
(a) surface, (b) flush, (c) plug with cover removed.

Protection of these socket-outlet circuits is by means of sub-circuit fuses of appropriate rating, bearing in mind that the current rating of fuse must not exceed the current rating of the cable.

57

Ring Circuits

One of the disadvantages of the radial circuit for socket outlets is the tendency to feed a number of appliances with the attendant risks of trailing flexes.

A more modern system of connecting socket outlets is by a ring circuit (Fig. 9.3). In this the wiring, which must be 2·5 mm² if rubber or

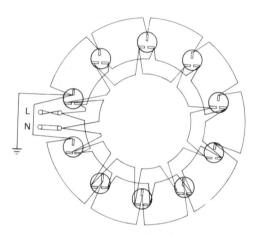

FIG. 9.3. Socket outlets on ring circuit.

p.v.c. insulated and 1·5 mm² if mineral-insulated takes the form of a ring. Both ends of the live conductor are taken into the terminal of a 30 A fuse, both ends of the neutral into the neutral terminal, and both ends of the earth wire into the earthing terminal, except when metal conduit is used. Thirteen ampere socket outlets having fused plugs may be connected to the ring. For certain purposes (e.g. the connection of stationary appliances such as panel fires) a fused spur box is used.

It is permissible to feed 13 A socket outlets by means of spurs taken from a ring circuit provided that the total number of spurs is not greater than the number of socket outlets and stationary appliances on the ring (Fig. 9.4). The spurs can be taken either from a socket outlet point box, from the origin of the ring, or from a fused spur box. No more than two socket outlets or one stationary appliance can be connected to any one non-fused spur.

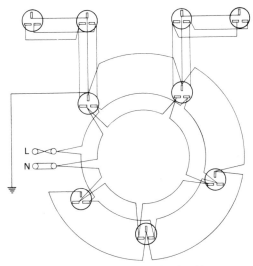

Fig. 9.4. Socket outlets on ring circuit with spurs.

A distinction is made in the *IEE Regulations* between the use of 13 A socket outlets (Fig. 9.5) in domestic and non-domestic premises. In domestic premises a ring final subcircuit must not serve an area greater than the extent of 100 m² of floor area. In industrial, commercial, and other non-domestic premises the requirement is based on the condition that owing to diversity, the maximum demand of apparatus is unlikely to exceed the maximum permissible fuse rating.

Socket outlets of more than 15 A rating are sometimes used in non-domestic premises. These are usually of a type designed for use with two-pin and earth-contact plugs, with single-pole fusing on the live pole. They may be connected to either radial or ring final subcircuits protected by a fuse of 30 A maximum capacity in much the same way as 13 A sockets, but in this case the number of socket outlets on a subcircuit must be such that the loading of the subcircuit does not normally exceed 30 A. The use of non-fused spurs is not permissible in this case.

If connected to a d.c. supply, a socket outlet, whatever its capacity, must be controlled by a switch which can be either adjacent or combined with the socket outlet. Socket outlets connected directly to mains supplies must not be installed in bathrooms. For electric shavers, a point isolated by a double-wound transformer unit complying with BS 3052 may be used in a bathroom.

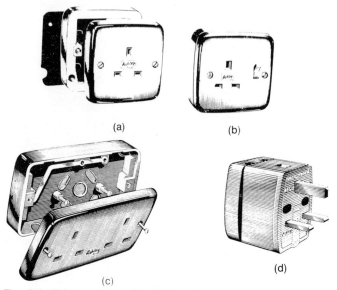

(a)   (b)

(c)

(d)

FIG. 9.5. Thirteen-amp socket outlets: (a) unswitched, (b) switched, (c) twin type, (d) 13 A fused plug.

Connecting socket outlets

When connecting socket outlets to radial circuits it is usual, after stripping sufficient insulation to allow the conductors to enter the terminal, to twist the conductors together (when there is more than one). In the case of socket outlets or ring final subcircuits the conductor is sometimes looped into the terminal without cutting. If the conductor is cut, the joint must be such that electrical continuity is ensured.

Connecting Plugs

The connection of a plug to the flexible cord or cable of an appliance should be carried out with great care. Many faults have resulted from neglect in this respect. Important points to bear in mind, with reference to Fig. 9.6, are:

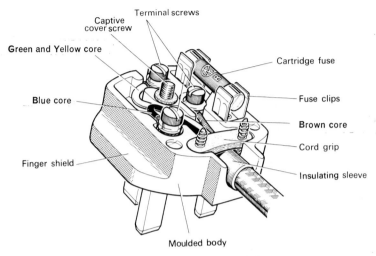

FIG. 9.6. Detail of 13 A fused plug.

(a) The sheath and braiding (if any) of the flexible should be removed for a sufficient distance to permit proper entry of the conductors into the terminals while leaving enough of the sheath to enter beneath the cord grip (an insulating sheath should preferably be slipped over the sheath at this point).
(b) The ends of the cores should be stripped for the correct distance so that when laid in the grooves in the plug the insulation just touches the terminal in each case, and the bare conductor just emerges from the terminal.
(c) the correct connections should be made (the live brown-core conductor should be connected to the fused terminal on the right-hand side when viewed from the back; the neutral blue-core conductor to the left-hand terminal; the earth green-and-yellow-core conductor to the top centre terminal).
(d) The terminal screws should be screwed home tightly, remembering that a loose connection causes overheating, and the

screws of the cord grip should be tightened just sufficiently to grip the end of the sheath firmly.

Spurs

A spur is really a branch cable connected to a ring circuit. Non-fused spurs may be installed in domestic premises provided that the cable conductor has a current rating not less than that of the conductors forming the ring. Not more than two single socket outlets, one twin socket outlet, or one stationary appliance must be fed from a non-fused spur.

Fused spur boxes (Fig. 9.7), which may be either switched or un-switched, provide a satisfactory means of connecting a spur to a ring. The rating of the fuse within the spur box should not exceed the rating of the cable forming the spur, therefore a 1·5 mm² cable must be protected by a 13 A fuse. In any case the fuse rating should never be more than 13 A.

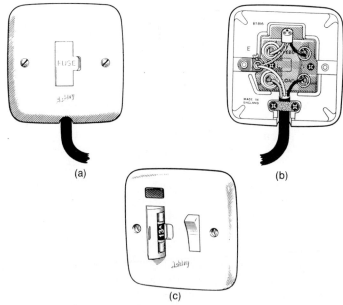

FIG. 9.7. Fused spur boxes: (a) front view, (b) rear view, (c) switched type, showing carrier pivoted for fuse replacement.

Questionnaire No. 9

1. A socket outlet and plug may be considered as two portions of a device for....
2. Older-type radial circuits use ... plugs and sockets of ... capacity.
3. In a ring circuit, both ends of the live conductor are taken into ....
4. It is permissible to feed 13 A socket outlets by means of spurs taken from a ring circuit, provided that ....
5. In domestic premises, a ring final subcircuit must not serve an area greater than ....
6. Socket outlets of more than 15 A rating are sometimes used in ....
7. If connected to a d.c. supply, a socket outlet must be ....
8. When connecting a plug to a flexible cord, special care should be taken in: (a) ..., (b) ..., (c) ..., (d) ....
9. A spur is ....
10. Non-fused spurs may be installed in ... provided that ....

Lighting Subcircuits

To control one or more lighting points from one position, a single-pole
one-way switch is normally used.

BEFORE an electric current can flow there must be a complete path,
or circuit. In other words, there must be an arrangement of conductors to carry the current from and to the source of supply. These
conductors forming the circuit include, beside cables, fuses, and

circuit-breakers, switches and items of equipment such as lighting points, which are to be connected to the supply.

A circuit from a distribution board designed to supply electrical energy directly to current-using apparatus is known as a final subcircuit.

Probably the simplest type of circuit for lighting is the series circuit used for Christmas-tree decoration lights. In this a number of lamps are connected together by conductors so as to form one complete, continuous path. The supply voltage is therefore shared by all of the lamps. Thus, on a 240 V supply, for example, there could be twenty 12 V lamps or fifteen 16 V lamps.

The type of circuit is not used in ordinary general service lighting for the following reasons:

(a) Unless some form of automatic shorting device is provided for each lamp, failure of any lamp will stop the flow of current and cause all of the lamps to go out.

(b) It would not be practicable to control lamps individually by switches.

(c) To permit any variation in the number of lamps on a circuit, lamps of many voltages would have to be made available, which would be inconvenient and unsatisfactory.

In practical wiring, the arrangement of a circuit depends to some extent on the wiring system employed.

To some extent also, the circuit will be affected by the method of wiring. This term, when used in connection with lighting subcircuits, indicates the way in which connections are made between adjacent lights and switches. Three common methods already described are: (1) looping-in; (2) joint-box; (3) three-plate ceiling rose. Looping-in is used for single-core cables, the joint-box method for two-core or multi-core cables and three-plate ceiling roses for either single-core, two-core, or multi-core cables.

Circuit conductors of the systems and methods previously mentioned are identified by coloured insulation (Fig. 10.1). In the case of installations connected to two-wire supplies (e.g. live and neutral), conductors of single-core cables between a live terminal and the lighting joint are provided with red insulation and conductors between a neutral terminal and the lighting point with black. This coloured insulation does not always extend the whole length of the conductor. In the case of mineral-insulated cables, for instance, or when a cable having a red and black core is taken to a switch, the identification takes the form of a red tape or sleeve at the termination.

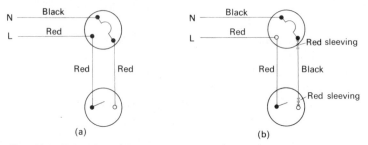

Fig. 10.1. Colour identification of: (a) single-core cables, (b) terminations of insulating sheathed cables.

Control

Lighting points are controlled by connecting switches in the circuits which-feed or supply the points. Circuits can be controlled by either single-pole switches, which are designed to open or close on one pole or phase only, or linked switches, which break all poles or phases at the same time or in a definite order.

To control one or more lighting points independently from one position, a single pole one-way switch is normally used (Fig. 10.2). Only a limited form of control can be obtained by connecting two one-way switches in a circuit.

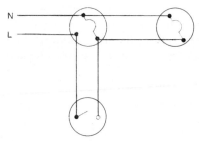

Fig. 10.2. Two lighting points controlled by one one-way switch.

If it is required to control a lighting point or a number of lighting points independently from two positions (for instance a staircase light so arranged that it can be switched "on" or "off" from either the top or bottom of the stairs) two two-way switches are used (Fig. 10.3). There are two alternative ways of connecting switches to give two-way control.

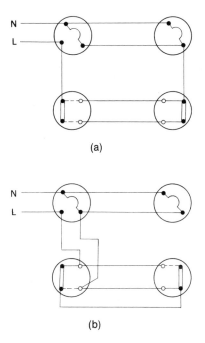

(a)

(b)

FIG. 10.3. Two lighting points controlled by two two-way switches: (a) conventional circuit, (b) alternative circuit.

In cases where control from three or more positions has to be provided, two-way switches are used in conjunction with intermediate switches (Fig. 10.4). Any number of intermediate switches can be inserted between the two-way switches to give independent control from each switch position.

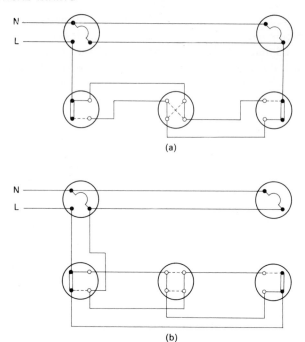

(a)

(b)

FIG. 10.4. Two lighting points controlled by two two-way switches and one intermediate switch: (a) conventional circuit using type *a* intermediate switch, (b) alternative circuit using type *b* intermediate switch.

Questionnaire No. 10

1. Before an electric current can flow, there must be . . . .
2. A final subcircuit is . . . .
3. Series circuits are not used for ordinary general service lighting because: (a) . . ., (b) . . ., (c) . . . .
4. Wiring systems can be grouped into: (a) . . ., and (b) . . . .
5. The looping-in method of wiring is used for . . . .
6. The joint-box method of wiring is used for . . . .
7. The three-plate ceiling-rose method of wiring is used for . . . .
8. Insulation of single core cable conductors between a lighting point and (a) a live terminal is coloured . . ., and (b) a neutral terminal is coloured . . . .
9. Single-pole switches are designed to . . . and double-pole switches to . . . .
10. A one-way switch is used to . . . and two two-way switches to . . . .

Lighting Accessories

If lampholders are used for flexible pendants in which the cord
supports the lamp, they must have an effective cord grip.

AN ACCESSORY is defined as "any device, other than a lighting fitting,
associated with the wiring and current-using appliances of an instal-

lation". Accessories used in conjunction with electric lighting include ceiling roses, lampholders, switches, and joint boxes.

Ceiling Roses

A ceiling rose is normally situated at a lighting point, that is at the termination of the fixed wiring. It provides a safe and convenient means of supporting a flexible cord pendant and jointing it to cable conductors. It consists basically of brass terminals mounted on or in an insulating base, provided with an insulated cover, and incorporating a grip for flexible cord.

The ceiling rose (Fig. 11.1) may have either two, three, or four terminals. In the three- or four-terminal patterns one of the terminals, the "loop-in" terminal, is shrouded with insulating material.

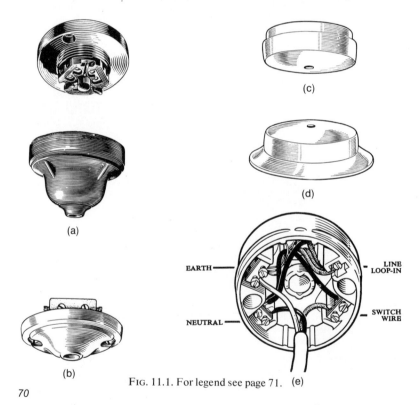

(c)

(d)

(a)

(b)

FIG. 11.1. For legend see page 71. (e)

EARTH

NEUTRAL

LINE LOOP-IN

SWITCH WIRE

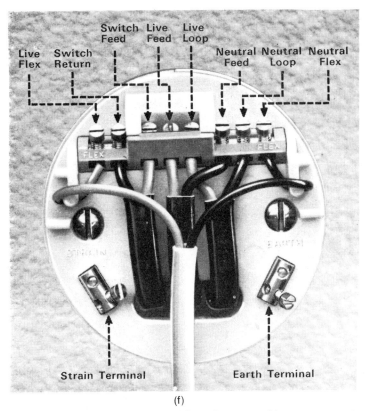

(f)

Fig. 11.1. Ceiling roses: (a) porcelain surface type with cover removed, (b) porcelain semi-recessed type, (c) moulded surface type with integral backplate, (d) moulded surface "halo" type with integral backplate, (e) four-terminal ceiling rose showing connections, (f) advanced design with strain terminal and shrouded live terminal.

This is because a loop-in terminal will probably be live even when the switch controlling the point is in the "off" position and must not, therefore, be accessible. One terminal is required for earthing.

Ceiling roses can be arranged for mounting direct on building surfaces, conduit boxes, or insulating pattresses. They should permit the cable or core terminations to be enclosed in incombustible material.

Lampholders

As the name implies, these are used to hold lamps. They also keep lamps in contact with the circuit conductors. Mains voltage filament lamps rated up to and including 150 W have caps provided with bayonet contacts; above this rating screwed caps are used. Naturally, the lampholders are designed accordingly.

Bayonet-cap lampholders (Fig. 11.2) are usually of insulated exterior and provided with spring-loaded brass contact pins. The prongs of the lamp cap fit into slots in the side of the lampholders, which also serve to maintain the lamp contacts pressed against the pins.

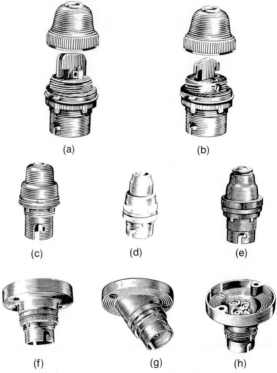

(a) (b)

(c) (d) (e)

(f) (g) (h)

FIG. 11.2. Bayonet-cap lampholders: (a) solid plunger, (b) spring plunger, (c) cord grip, (d) threaded, (e) TRS gland, (f) standard batten, (g) angle batten, (h) loop-in batten.

Although lampholders are sometimes fixed direct to the surfaces of ceiling or walls ("batten" type) they are mostly designed for fixing to flexible cord. If used for flexible pendants in which the cord supports the lamp, they must have an effective cord grip. When used with TRS or PVC-sheathed flexible cord they have glands to grip the sheath. If used for lighting fittings involving bracket fixing or tube suspension, they are provided with threaded entry. In certain circumstances they are required to be heat-resistant.

For securing shades, threaded shade rings are screwed on the outside of the lampholder body. Alternatively, short "skirts" are used, although Home Office type ventilated safety shields provide more protection.

Switches

We have already seen from our study of circuits that a switch is used for closing or opening a circuit or part of a circuit. Switches which control subcircuits, often referred to as "subswitches", may be operated by the up and down movement of a dolley or rocking bar, by pulling a cord, by rotating a switch knob, or by some other means (Fig. 11.3).

The conducting portions of a switch are the terminals and the contact surfaces. These are mounted in or on insulating material. The operating mechanism, which brings together and separates the contact surfaces, often depends on the action of springs.

A single-pole one-way switch normally has two terminals and a mechanism which provides make and break. In other words, in the "off" position of the switch the contacts provide a gap in the circuit which is closed when the switch is in the "on" position.

A single-pole two-way switch has three terminals; one is often called the "bar" terminal and the other two the "strapping" terminals, although they are not necessarily connected in the same way. There is, of course, no fixed "on" and "off" position of a two-way switch.

Single-pole intermediate switches have four terminals. Two different patterns are in common use — one requiring horizontal connection of strapping wire with one crossed over and the other requiring straightforward vertical connection. As with two-way switches, there is no definite "on" or "off" position of an intermediate switch.

Besides two-way and intermediate subswitches, there are various special types of manual switch such as two-way and off, series-parallel, reversing, and inductive circuit. Automatic types include circuit-breakers, thermostats, clock switches, and discharge-lamp starters.

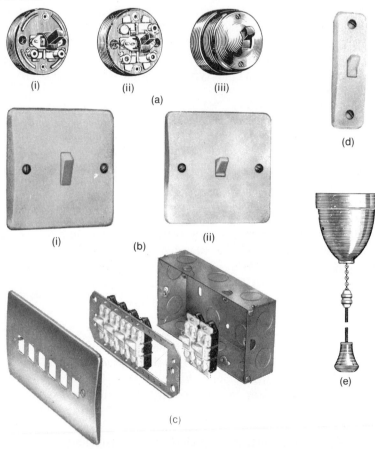

Fig. 11.3. Switches: (a) round surface type; (i) one-way with cover removed, (ii) two-way with cover removed, (iii) complete. (b) Flush type; (i) with rocker, (ii) with dolly. (c) Grid type. (d) Architrave type. (e) Cord-pull ceiling switch.

A particular application of series-parallel control is the rotary three-heat switch (Fig. 11.4) used to control a two-section heating element so that three different heat levels can be obtained. These are available as both single-pole (four terminals) and double-pole (five-terminals).

Fig. 11.4. Rotary switch.

Reversing switches are used to reverse small motors, demagnetize magnets, and to change the polarity of d.c. circuits used for fluorescent lighting. Inductive circuit switches introduce resistance into an a.c. circuit containing coils to prevent damage by induced effects when the break occurs.

Circuit-breakers have already been mentioned as an alternative to fuse protection. They are also utilized for control purposes. Often they can be operated by simply pressing push buttons which open or close contacts in the coil circuit (Fig. 11.5).

By means of a thermostat, it is possible to maintain automatically at a fairly constant level the temperature of the air or water surrounding it.

An arrangement incorporating a thermostat with its own built-in heater enables several alternative heat energy levels to be obtained from an element.

Clock switches, or time switches, consist of clocks coupled to contacts so that at preselected times the circuits they control are switched "on" or "off". They are commonly employed for shop-window or sign lighting, to control off-peak storage heating, electric cooking, etc.

FIG. 11.5. Push-button switch.

The purpose of fluorescent lamp starter switches (Fig. 11.6) is to cause a sudden interruption of current so that a voltage surge sufficient to start the discharge across the lamp will be created. There are two main types: (1) thermal, having four terminals, and (2) glow type, having two terminals.

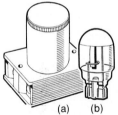

(a)   (b)

FIG. 11.6. Fluorescent lamp starter switches: (a) thermal, (b) glow.

In the thermal type two contacts which are normally closed are operated by the heat from a small element mounted adjacent to them. In the glow type there are two normally open contacts which are caused to touch together as a result of the heat from a glow discharge and then to spring apart when the discharge ceases.

The ordinary manually operated main switches and switch fuses are available enclosed in steel cases with interlocked covers or as all-insulated units. To comply with the *IEE Regulations* there must be provision for totally enclosing cable terminations in incombustible material.

Joint Boxes

These vary according to the specific purpose for which they are intended. Those for containing joints in TRS or PVC-sheathed cables (Fig. 11.7) usually consist of brass terminals fixed in boxes of moulded insulating material and provided with lids. They may contain three, four, five, or six terminals. Alternatively, they may have no terminals at all and may be designed to contain connectors.

(a)

(b)

Fig. 11.7. Joint boxes: (a) for insulating sheathed cables, (b) for metal-sheathed cables.

The terminals of boxes used for ring circuits should, of course, be of at least 30 A rating and should be capable of accommodating four 2·5 mm² conductors.

Metal joint boxes (see Fig. 11.7) may be used in conjunction with metal-sheathed or armoured cables in which the sheath or armour is used as an earth-continuity conductor. However, if non-metallic joint boxes are used with these cables there must be an effective method of maintaining the earth continuity. This can be achieved by a metal bonding strip having a resistance no higher than that of an equivalent length of earth-continuity conductor.

Questionnaire No. 11

1. An accessory is defined as . . . .
2. A ceiling rose consists basically of . . . .
3. The loop-in terminal is shrouded because . . . .
4. Bayonet-cap lampholders usually consist of . . . .
5. A switch is used for . . . .
6. A one-way switch has . . . terminals, a two-way switch . . . terminals, and an intermediate switch . . . terminals.
7. Besides two-way and intermediate subswitches, there are various special types of manual switch such as (1) . . ., (2) . . ., (3) . . ., (4) . . ..
8. Automatic switches include: (1) . . ., (2) . . ., (3) . . ., (4) . . ..
9. The two types of fluorescent lamp starter switches are: (1) . . ., (2) . . . .
10. Joint boxes for use with TRS or PVC-sheathed cables usually consist of . . . .

Bells

A bell is used to give an audible alarm.

A BELL is used to give an audible alarm or a signal. That is either to engage attention or convey a message. The simple house door-bell, for instance, is used to inform an occupier that someone outside wishes either to say something or to gain admittance. A person hearing a fire or burglar alarm bell knows that he should immediately inform the fire brigade or the police.

Principle

Electric bells are generally based on the principle of the electro-magnet, in other words, they depend on magnetism produced in

79

a soft iron core by the flow of electric current through a coil surrounding the core. In most bells there are two coils arranged side
by side each containing one leg of a common core (Fig. 12.1).

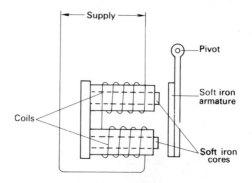

Fig. 12.1. Principle of electromagnet.

The electromagnet cores are arranged in a suitable position close
to a pivoted iron armature so that when current flows in the coils
and the cores become magnetized, they attract the armature towards
them. A striker, or hammer, is attached to the armature so that when
the attraction takes place the hammer is caused to hit a nearby gong
and result in sound. When current ceases to flow, the core becomes
demagnetized and, due to the action of a spring, the armature
returns to its original position.

A bell is usually controlled by a push connected in series in one
of the supply leads.

Types

Electric bells are made with various shapes of gong. There are
round gongs, sheep gongs, church gongs, and wire gongs, each
causing a distinctive sound (Fig. 12.2). Underdome bells (Fig. 12.3)
are designed so that their hammers strike the inside surface of the
domes and, as the coils are contained within the domes, occupy
less space than the original pattern. Buzzers, which have no gong
at all and rely solely on a vibrating member for their sound, are
sometimes used as alternatives to bells.

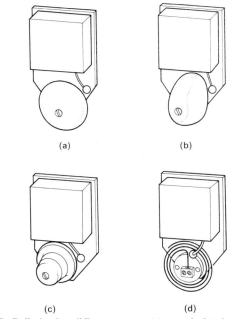

(a)   (b)

(c)   (d)

FIG. 12.2. Bells having different gongs: (a) round, (b) sheep, (c) church, (d) wire.

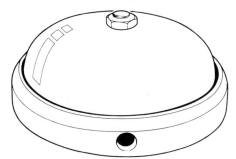

FIG. 12.3. Underdome bell.

Most door-bells are operated at extra-low voltage, usually obtained from a transformer connected to the a.c. mains supply. However, for some purposes, such as outdoor burglar and fire alarms, bells which operate at mains voltage are used.

Single-stroke Bell

This is the simplest kind of electric bell. In the standard pattern the coils are generally mounted on the base of the bell and the electric supply is taken via a push to the two terminals (Fig. 12.4).

When the push is pressed the circuit is completed and current flows through the coils magnetizing the cores within them so that they attract the armature and cause the hammer to strike the gong once. As soon as the push is released, the armature is restored.

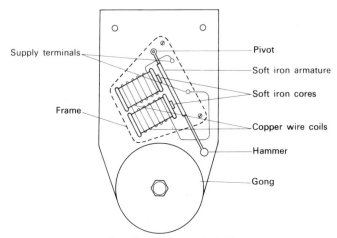

Supply terminals
Pivot
Soft iron armature
Soft iron cores
Frame
Copper wire coils
Hammer
Gong

FIG. 12.4. Single-stroke bell.

Because this arrangement produces sound for only a very short time, it is unsuitable for use as a warning bell. Its principal function is for signalling. For example, a bell on a public transport vehicle can be rung once to instruct the driver to stop the bus and twice to start it.

Trembling Bells

Ordinary door-bells are of the trembling type. While the circuit is made, the hammer makes a series of blows on the gong. This is achieved by introducing a make and break contact between the coils and one of the terminals (Fig. 12.5).

The contact is normally made so that, on pressing the push the armature moves towards the electromagnet against the action of a spring and the hammer strikes the gong. However, as the armature

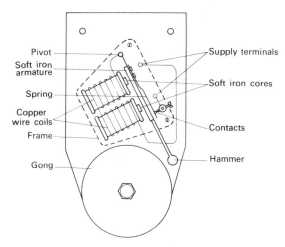

FIG. 12.5. Trembling bell.

moves forward the contact is broken and current stops flowing. Therefore, the magnetism ceases and the spring causes the armature to return to its original position.

As the armature is restored the contact is remade. The sequence of operations is repeated over and over again as long as the push is pressed. In other words the armature adopts a trembling motion, which gives the bell its name.

There is usually an adjustment screw by which the gap between the contacts can be adjusted. Sparking between the contacts faces is liable to lead in time to oxidation of the metal which renders the contact imperfect. In some bells the contact faces are of silver or platinum as these metals do not oxidize very readily. Trembling bells for use in situations where explosives or flammable gases may be present are of a special flame-proof pattern.

Two or more trembling bells will generally operate satisfactorily in parallel but not in series. This is because series connection would complicate the switching action. The usual way of operating in series is to connect one or more single-stroke bells in circuit with one trembling bell so that the one make-and-break results in the trembling action of all of the armatures.

Continuous-ringing Bells

For certain purposes (for instance fire and burglar alarms), it is desirable to have a bell which, once started, will continue to ring even after the push or contact is released. This state of affairs can be obtained either by including a special kind of control or relay in a trembling bell circuit, or by using a continuous-ringing bell (Fig. 12.6).

The continuous-ringing type of bell has three terminals. Two are connected to the bell electromagnet coils via contacts in the usual way; the third terminal is connected to a second set of contacts. In some cases there is a separate electromagnet to operate the second contacts.

In addition to the normal trembling bell arrangement, this type of bell has a spring-supported trigger and catch to operate the second set of contacts. Normally, the trigger rests on the catch which projects from the armature. As soon as the armature moves forward to start its trembling action the trigger, which is pivoted, is released from the catch. The trigger carries one of the contact faces and, due to the spring, the two contact faces are brought together. The making of this contact shorts out the push, and the bell is then connected directly to the supply. To stop the bell from ringing, a resetting cord is pulled to restore the trigger to its horizontal position resting on the catch.

Magneto Bells

For telephone purposes, a trembling bell has certain disadvantages. The breaking at the contacts causes a high induced e.m.f. and when the bell is in frequent use this involves renewal of the contacts. Therefore magneto bells, which operate from alternating current and entail no make and break, are used (Fig. 12.7). Due to the constantly changing polarity the armature is attracted during each half-cycle. To improve the sensitivity, the bell is polarized by the introduction of a permanent magnet, so that during one half-cycle the flux due to the alternating current assists that of the

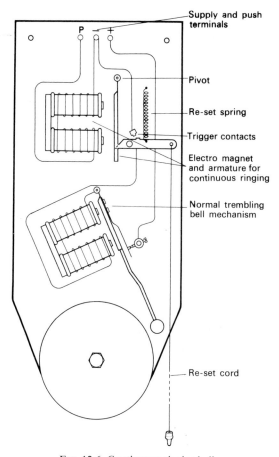

FIG. 12.6. Continuous-ringing bell.

permanent magnet and causes attraction in the armature, and during the other half-cycle the current flux opposes the permanent magnet flux and does not attract. Thus, in this case, the armature is attracted only once per cycle. The hammer is situated between two gongs, so that movement of the armature causes each gong to be struck alternately.

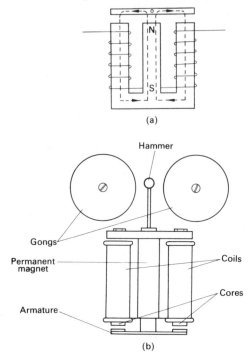

FIG. 12.7. Magneto bell: (a) principle, (b) arrangement.

Chimes

A more modern trend in warning sounds is for homes to be equipped with electric door chimes instead of bells. The simple "double-note" type depends for its action on a solenoid plunger pulling against a spring to act as a striker. Two tubes of different lengths provide the two notes.

In a more expensive pattern, the familiar first eight notes of the Westminster chimes are played.

In the usual pattern of door chimes a core within a coil is attached to a spring and provided with two nylon strikers. The strikers are arranged one at each end of the core so that sideways movement causes each to knock against a tube at the side.

When the front-door push is pressed, the core is caused to move and one of the nylon strikers hits one of the tubes. As soon as the push is released, the core moves back sharply under the action of the spring and the striker at the other end of the core strikes the other tube. Thus, pressing the push produces the characteristic two-note chime.

If the coil is also connected to the supply via a second push situated at the back door (Fig. 12.8) with a suitable value of resistance in series, the movement of the striker against the first tube is

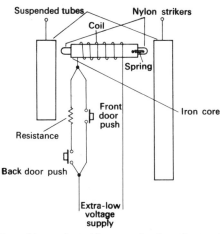

Fig. 12.8. Door chimes adapted for operation from front and back doors.

less sharp than before and spring action is insufficient to cause striking against the second tube. The householder can therefore tell which push has been pressed according to whether the sound consists of one note, or two notes.

Bell Pushes

The principle of an ordinary push is simple. Two contacts are closed when a spring is depressed by pushing a plunger. There are a variety of patterns: desk, barrel, pear, torpedo (Fig. 12.9). The first two can be obtained as single or multi-way units, surface or flush, indoor or watertight.

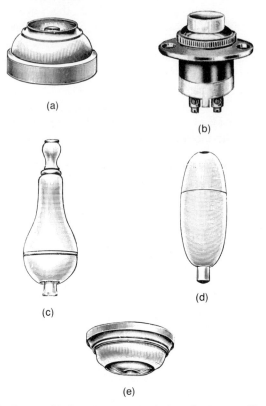

FIG. 12.9. Types of bell push: (a) desk, (b) barrel, (c) pear, (d) torpedo; (e) bell rosette.

Pear and torpedo pushes are suspended on flexible cords from a rosette. Bell rosettes, which are the equivalent of ceiling roses in lighting circuits, are made with up to twelve plates.

For certain special purposes (e.g. for signalling) two-way or "morse-key" pushes are used. These have three contacts; two are normally closed and an alternative connection is made when the push is depressed.

When an "off" position must be provided, bell switches are required. They are available with from one to six ways.

Bell Indicators

When one bell can be operated by two or more pushes, for example in a hotel, it is necessary to have some method of indicating to the person being called which push has been pressed. The simplest types of indicator board have as many ways as there are pushes. Each way is connected in series with a push so that pressing the push operates that particular way as well as the bell and gives a visual indication as to where the call originated.

The indication may take the form of a flag appearing in a small window or of the window being illuminated.

The portion that causes operation of a flag indicator way is known as the movement. Three common types of movement are: (1) pendulum, (2) mechanical replacement, (3) electrical replacement. In common with the electric bell, the basis of the movement is an electromagnet.

In the pendulum movement (Fig. 12.10), an armature carrying the

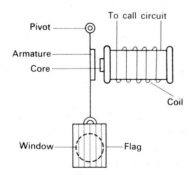

FIG. 12.10. Pendulum indicator movement.

flag is pivoted close to the electromagnet, the coil of which carries current when the push in series with it is pressed. The armature is attracted towards the core within the coil and as soon as the push is released, the flag swings in the window. Two disadvantages of this type of indicator are: (a) if the person called does not happen to be

89

near to the indicator when the bell rings, the flag may have stopped swinging by the time he or she reaches the indicator, (b) it would not be suitable where vibration may be set up.

A mechanical replacement movement (Fig. 12.11) does not have

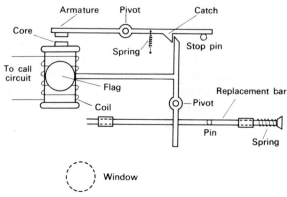

FIG. 12.11. Mechanical replacement indicator movement.

these two disadvantages. When the bell push is pressed the electromagnet releases a catch and the flag moves into the window. In this case the flag does not swing but remains in its new position. After the call has been noted, the flag is restored to its original position by manual operation of a mechanical resetting device. Once restored, the catch retains the movement until the electromagnet is energized again by pressing a push.

In the third type of movement, the resetting or replacing is carried out by an electromagnet operating in the reverse direction to the electromagnet which causes the flag to appear in the window. In this case there is one more push than there are indicator ways, the additional push being used to control the electromagnets which reset the movements. The coils of the resetting electromagnets are usually connected in series.

There are two main patterns (Fig. 12.12) of electrical replacement movement, known as: (1) polarized, having operating and resetting coils wound on a common core and including a permanent magnet, and (2) non-polarized, having separate electromagnet for calling and resetting. The first type, since they depend on permanent magnetism, are suitable for operation on d.c. supplies only. The non-polarized pattern can be operated from either d.c. or a.c. supplies.

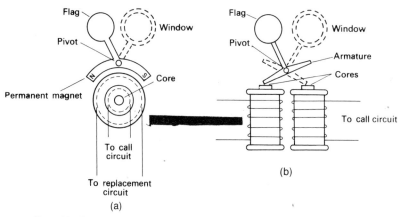

FIG. 12.12. Electrical replacement indicator movements: (a) polarized type, (b) non-polarized type.

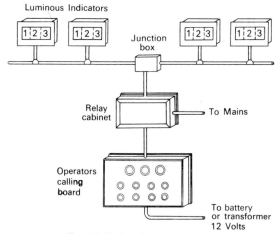

FIG. 12.13. Luminous call system.

91

All three types of indicator can be obtained for either surface flush or semi-flush mounting. There are also specially housed indicators for tropical and marine use, etc.

The "silent call" (luminous) system (Fig. 12.13) is used in hospitals and nursing homes, where the minimum audible signal is required. In a typical installation of this nature, the pressing of a ward-calling push operates a buzzer in the nurses' duty room, and also lights up the appropriate way of the indicator and a lamp at the ward door. On releasing the ward push, the buzzer ceases but both indicator lamps remain "on" until the call has been answered and a reset push in the ward pressed.

Luminous indicators are also used for "engaged/enter" signs and lights.

Bell Circuits and Wiring

There are many practical circuits in which one or more bells are controlled from one or more positions, with or without indicators.

Bell circuits operated at extra-low voltage are usually wired in 0·75 mm² copper wire insulated with PVC, rubber, or waxed cotton. This is available in either single or flat twin form and in a variety of colours to make for easy identification (there are no regulations as to colours of cables for extra-low voltage circuits) (Fig. 12.14).

Standard bell flexibles are copper, double silk, or cotton covered, twisted and braided. Normally this is in two, four, six, ten, and fifteen cores. Rubber- and plastic-covered 0·5 mm² copper bell flexible can also be obtained.

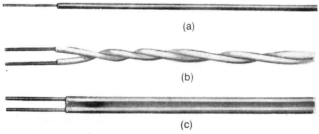

(a)

(b)

(c)

Fig. 12.14. Bell wires: (a) single, (b) twin twisted, (c) figure-8.

When run on suitable surfaces (e.g. wood), extra-low voltage wiring can be fixed by fibre-insulated, coppered-steel staples. Alternatively, it may be run in steel conduits sunk in floors, walls, etc. (not, however, bunched in the same conduit as low-voltage wiring).

Junction boxes for use with this type of bell wiring are made with up to twenty connection plates.

Low-voltage mains bell wiring must comply in all respects with the *IEE Regulations*; this also applies to wiring feeding a transformer from which extra-low voltage bells circuits are supplied.

Bell Transformers

If two coils are wound on the same iron core and one coil is connected to the a.c. mains, the voltage across the other coil will depend on the ratio of the number of turns of the coils. For example, if one coil with 480 turns is connected to 240 V a.c. and the second coil has 12 turns, the voltage across this second coil will be

$$240 \times \frac{12}{480} = 240 \times \frac{1}{40} = 6 \text{ V}.$$

This arrangement of two coils wound on a common core (Fig. 12.15) is known as a double-wound transformer. When it is used to reduce the voltage, as in the case of an extra-low voltage bell circuit, it is said to "step down" the voltage. To give some variation in output voltage, it is usual to take "tappings" from the second or "secondary" coil. If these tappings are taken so that there are eight and sixteen effective turns making up a total of twenty-four the secondary voltages obtainable are:

(1) First 8 turns: $240 \times \dfrac{8}{480} = 240 \times \dfrac{1}{60} = 4 \text{ V}.$

(2) Last 16 turns: $240 \times \dfrac{16}{480} = 240 \times \dfrac{1}{30} = 8 \text{ V}.$

(3) All 24 turns: $240 \times \dfrac{24}{480} = 240 \times \dfrac{1}{20} = 12 \text{ V}.$

The *IEE Regulations* permit a bell transformer to be supplied either from a separate way of a fuseboard or from an existing subcircuit. If from an existing subcircuit, the transformer should preferably be fed from an adjacent socket outlet. In addition, to comply with the *IEE Regulations*, one point of the secondary winding and the metal core and housing (unless double-insulated) must be earthed. The best bell transformers are fused on both the mains voltage and the secondary sides.

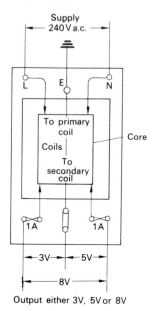

Output either 3V, 5V or 8V

FIG. 12.15. Arrangement of bell transformer.

Questionnaire No. 12

1. Electric bells are generally based on the principle of . . . .
2. The principal function of a single-stroke bell is . . . .
3. The action of a trembling bell is as follows . . . .
4. A continuous-ringing bell is used for purposes such as . . . .
5. For telephone purposes . . . bells, which operate . . . are used.
6. There are a variety of pushes, including (1) . . ., (2) . . ., (3) . . ., (4) . . . .
7. The purpose of a bell indicator is . . . .
8. Three common types of bell indicator movement are: (1) . . ., (2) . . ., (3) . . . .
9. Extra-low voltage bell circuits are usually wired in . . . .
10. To obtain 12 V from a bell transformer with 800 primary turns connected to a 240 V supply, a secondary with . . . turns would be required.

Primary and Secondary Cells

Electricity can be produced as a result of chemical activity.

ELECTRICITY can be obtained in various ways. In the case of most public supplies, each generator consists of a large number of conductors rotating in a magnetic field. The output of such generators is

95

alternating current, the voltage of which can be readily changed by using a transformer. For electric-bell circuits, as we have already seen, the voltage is generally reduced from 240 to 12 or less.

Years ago, when public supplies were direct current, which cannot easily be transformed to a lower voltage, the electricity supply for the majority of bells used to be obtained from cells, which depend on the fact that electricity can be produced as a result of chemical activity. Basically, a cell consists of two plates, or elements, immersed in a suitably active liquid. Two or more cells coupled together make up what is known as a battery.

Bells for certain purposes are still operated by batteries of cells, and a great many devices, particularly portable ones, rely on this type of electricity supply.

There are two main types of cells: (a) primary and (b) secondary. In a primary cell a chemical action generates electrical energy and when the chemical energy ceases the cell is thrown away. In a secondary cell, often called an accumulator, electrical energy is applied to produce a chemical change which will enable electrical energy to be drawn from the cell. In other words, the secondary cell is an electricity storage device. The process of applying electrical energy to such a cell, or to a battery of cells, is known as "charging".

Primary Cells

The commonest primary cell is that known as a Leclanché cell (Fig. 13.1). In its earliest form—the "wet" cell—it consists of a glass jar containing a zinc rod and a porous pot immersed in a solution of ammonium chloride in water (familiarly called sal ammoniac). In the centre of the earthenware porous pot is a carbon rod, which is the positive element, surrounded by a mixture of manganese dioxide and powdered carbon. The zinc rod is the negative element.

To appreciate how chemical energy is converted into electrical energy, it is best to think of an electric current as a flow of electrons, or negative particles. The chemical action of the sal ammoniac on the elements causes movement of electrons between the two elements, which results in the zinc becoming negatively charged and the carbon positively charged, so that if these two elements are

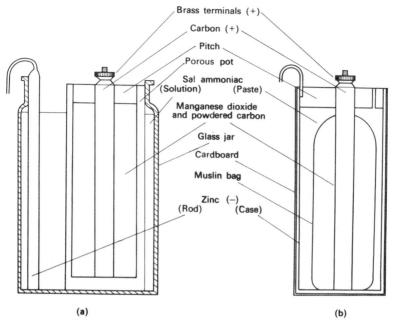

Brass terminals (+)
Carbon (+)
Pitch
Porous pot
Sal ammoniac
(Solution) (Paste)
Manganese dioxide
and powdered carbon
Glass jar
Cardboard
Muslin bag
Zinc (−)
(Rod) (Case)

(a) (b)

FIG. 13.1. Leclanché cells: (a) wet type, (b) dry type.

connected to an external circuit, a current flows from positive to negative.

If the positive element was immersed directly in the sal ammoniac, bubbles of hydrogen would collect on the surface of the carbon and spoil the action of the cell. This undesirable effect is called polarization. The manganese dioxide reduces this effect by releasing oxygen which combines with the hydrogen on the carbon rod to form water. Manganese dioxide is therefore called a depolarizer. The porous pot permits movement of liquids and gases, but prevents the powder within it from leaking out.

The zinc rod is coated with mercury amalgam which, while allowing the action of the cell to continue, prevents the zinc from rapidly dissolving (known as "local action").

A more familiar form of Leclanché cell is the so-called "dry" cell. This is not really a correct description, as no cell would operate unless liquid or electrolyte was present. The cell is made more portable by combining the active ingredients with zinc chloride, plaster of paris, and flour into a paste, placing the manganese dioxide in a sack and making the container of zinc so that it also functions as the negative element. The top is sealed with pitch, a vent hole being provided for gases to escape.

The Leclanché cell has an e.m.f. of about 1·5 V. The resistance of the wet cell is 1 Ω and of the dry cell 0·2 Ω. These cells are most suitable for intermittent use. If used to deliver current for long periods the manganese dioxide is unable to remove the hydrogen as quickly as it forms. The carbon rod then becomes covered with hydrogen bubbles which prevent effective contact, causing an increase in resistance and a corresponding reduction in e.m.f.

Other types of primary cells which are sometimes used in place of Leclanché dry cells are: the "air" cell, in which oxygen from the atmosphere takes the place of the manganese dioxide, and the "mercury oxide" cell, in which the positive element is a coil of corrugated zinc, the negative element mercury oxide and graphite, and the electrolyte potassium hydroxide.

There are also a number of cells in which the electrolyte is introduced immediately before the cell is to be used. This type has a comparatively short, active life.

The Gordon magnesium cell, used in hearing aids, has a carbon case for its positive element, a magnesium rod as the negative, and a weak solution of potassium bromide as the electrolyte. As this cell is only used to deliver very tiny currents (up to one-tenth of an ampere) no depolarizer is necessary.

Secondary Cells

The commonest of the secondary, or storage, cells is the lead–acid type. This has two plates, each consisting of a framework containing the active material. In the charged state the positive plate contains a brown material called lead peroxide, while the negative plate contains spongy lead. The electrolyte is sulphuric acid diluted with pure, or distilled, water.

During discharge there is a chemical action tending to turn the active material of both plates into lead sulphate, which reduces the strength, or specific gravity, of the diluted acid. The energy provided by this chemical action is converted into electrical energy.

However, long before both plates become lead sulphate the cell is recharged by connecting it to a direct current supply at the correct voltage. This process restores the positive to lead peroxide, the negative to spongy lead, and the specific gravity of the electrolyte to its original value. Gas is given off during charging, which should therefore be done in a ventilated room.

The characteristics of a lead–acid cell (Fig. 13.2) are as follows:

	Charged	Discharged
e.m.f. (V)	2·1	1·85
Specific gravity	1·21	1·17

Specific gravity is measured by using a hydrometer, an instrument into which a sample of the liquid to be tested is drawn so that a graduated float indicates its strength.

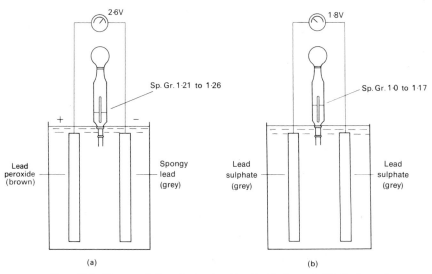

FIG. 13.2. Characteristics of a lead–acid cell: (a) charged, (b) discharged. (Note: voltage when charged falls rapidly to 2·1 when taken off charge.)

The normal discharge current multiplied by the time taken to discharge the cell at this current is called its ampere-hour capacity. This is always less than the number of ampere-hours needed to recharge it. The ratio of the two, expressed as a percentage, is the efficiency of the cell. Most lead–acid cells have an ampere-hour efficiency of about 85 per cent.

An alternative secondary cell to the lead–acid is the alkaline type. There are two types of these: (1) the Edison cell, having hydrated nickel peroxide as the positive element with finely divided iron as the negative, and (2) the Nife cell (Fig. 13.3), having a

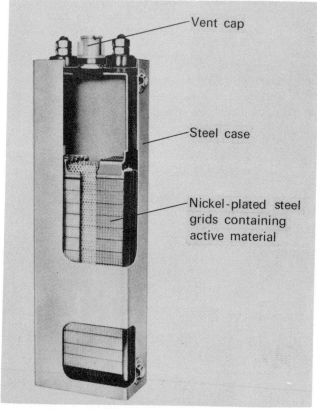

FIG. 13.3. Alkaline cell of nickel–cadmium type.

nickel hydroxide positive and a negative consisting of a mixture of cadmium and iron. Both types employ potassium hydroxide as the electrolyte. Fully charged, the cell has an e.m.f. of $1 \cdot 3$ V and discharged about $1 \cdot 1$ V.

Alkaline cells, although slightly dearer, have the advantage over lead–acid cells that they have longer life, are mechanically and electrically more robust (are not damaged by overcharging or shorting), and, since they can be completely sealed, are more easily stored.

A somewhat different type is the fuel cell. One kind has plates of porous nickel in an electrolyte of caustic soda solution. Oxygen is supplied to one plate and hydrogen to the other.

Charging Secondary Cells

To charge a secondary cell, direct current is passed through it in the opposite direction to that of the current obtained from the cell. Therefore the supply voltage must be above the fully charged voltage of the cell.

There are two recognized charging circuits (Fig. 13.4): (a) constant voltage, and (b) constant current.

In the constant voltage circuit used for small-scale charging (e.g. a car battery), a fixed value of resistance is connected in series with the cell or battery. As the charging progresses, the cell voltage increases and provides more and more opposition to the flow of current from the constant voltage d.c. supply, so that the current at the end of the charge is less than that at the start.

In the constant current method of charging, the resistance in series with the cells is varied, so that as the charge progresses the value of resistance is reduced. Thus the voltage actually applied after allowing for the cell voltage in the opposite direction is less at the end of the charge than at the start.

The normal rate of charge is influenced by the capacity of the cell or cells, which depends on the area of plate surface; the larger the surface, the greater is the capacity.

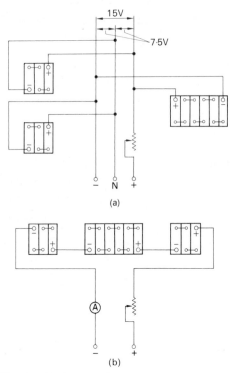

Fig. 13.4. Charging circuits: (a) constant voltage, (b) constant current.

Connecting Cells

To obtain a supply voltage greater than that of one cell, it is necessary to join a number of cells in series. For instance, three 1·5 V cells would provide 4·5 V. However, connecting three cells in series would mean that the total internal resistance would be threefold. The current that would be obtained from the battery of three cells in series (Fig. 13.5) would naturally depend on the resistance of the external circuit.

Supposing the battery of three 1·5 V cells in series, each with an internal resistance of 1 Ω, were connected to an external circuit of 0·5 Ω resistance. The current flowing is calculated as follows:

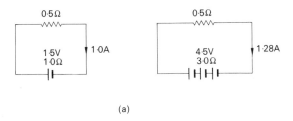

(a)

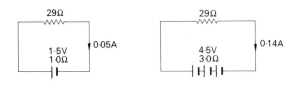

(b)

FIG. 13.5. Series connection of cells with: (a) 0·5 Ω external resistance, (b) 29 Ω external resistance.

Internal resistance $= 3 \times 1 \cdot 0 = 3 \cdot 0$ Ω.

Total resistance $= 3 \cdot 0 + 0 \cdot 5 = 3 \cdot 5$ Ω.

$$\text{Current} = \frac{\text{voltage}}{\text{resistance}} = \frac{4 \cdot 5}{3 \cdot 5} = 1 \cdot 28 \text{ A.}$$

With 1 cell only, the calculation would be:

Total resistance $= 1 \cdot 0 + 0 \cdot 5 = 1 \cdot 5$ Ω.

$$\text{Current} = \frac{\text{voltage}}{\text{resistance}} = \frac{1 \cdot 5}{1 \cdot 5} = 1 \cdot 0 \text{ A.}$$

This shows that in this case, although the voltage is three times as large, the gain in current is less than one-third.

Next, suppose the resistance of the external circuit were to be increased to 29 Ω. The current flowing when three cells were in series would become

$$\frac{4 \cdot 5}{32} = 0 \cdot 14 \text{ A,}$$

PRACTICAL WIRING

and with one cell only:

$$\frac{1 \cdot 5}{30} = 0 \cdot 05 \text{ A.}$$

Which indicates that nearly three times the current flows when the internal resistance is large in comparison with the external resistance.

When connecting cells in series it is very important that the positive terminal of one cell is joined to the negative terminal of the next cell. Any cell or cells wrongly connected would act in opposition to the remainder. If, in a circuit of three cells in series, one cell was in reverse, it would concel out the voltage of one of the other cells, leaving only one cell capable of being usefully employed.

It is, of course, possible to connect cells in parallel (Fig. 13.6). This does not alter the voltage obtained, but reduces the effective internal resistance.

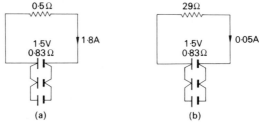

Fig. 13.6. Parallel connection of cells with: (a) $0 \cdot 5 \ \Omega$ external resistance, (b) 29 Ω external resistance.

Considering once again three 1·5 V cells, each of 1 Ω internal resistance. If connected in parallel with a $0 \cdot 5 \ \Omega$ resistance, the calculation is:

Internal resistance $= \dfrac{1}{3} = 0 \cdot 33 \ \Omega.$

Total resistance $= 0 \cdot 33 + 0 \cdot 5 = 0 \cdot 83 \ \Omega.$

Current $= \dfrac{\text{voltage}}{\text{resistance}} = \dfrac{1 \cdot 5}{0 \cdot 83} = 1 \cdot 8 \text{ A.}$

And when the external resistance was 29 Ω, the current would be

104

$$\frac{1 \cdot 5}{29 \cdot 33} = 0 \cdot 05 \text{ A}.$$

Comparing these two values of current with the 1 A and 0·05 A obtained when using only one cell, it is seen that the current gain is over three-quarters with the smaller current and zero with the larger.

Cells can be connected in series–parallel (Fig. 13.7). That is, a combination of series and parallel. For example, two cells could

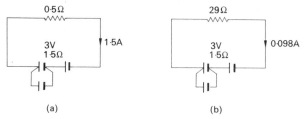

(a)   (b)

FIG. 13.7. Series–parallel connection of cells with: (a) 0·5 Ω external resistance, (b) 29 Ω external resistance.

be in parallel, then connected in series with a third cell. The calculation of current would then be as follows:

Voltage $= 1 \cdot 5 + 1 \cdot 5 = 3$ V.

Internal resistance of cells in parallel $= \dfrac{1}{2} = 0 \cdot 5$ Ω.

Total internal resistance $= 0 \cdot 5 + 1 \cdot 0 = 1 \cdot 5$ Ω.

Total resistance of circuit with 0·5 Ω external resistance
$$= 1 \cdot 5 + 0 \cdot 5 = 2 \text{ Ω}.$$

Current $= \dfrac{\text{voltage}}{\text{resistance}} = \dfrac{3}{2} = 1 \cdot 5$ A.

Total resistance of circuit with 29 Ω external resistance
$$= 1 \cdot 5 + 29 = 30 \cdot 5 \text{ Ω}.$$

Current $= \dfrac{\text{voltage}}{\text{resistance}} = \dfrac{3}{30 \cdot 5} = 0 \cdot 098$ A.

Questionnaire No. 13

1. A cell consists of . . . , and two or more cells coupled together make up what is known as . . . .
2. The difference between a primary cell and a secondary cell is . . . .
3. In a Leclanché cell the positive element is . . . , the negative element is . . . , and the liquid in which they are immersed is . . . .
4. Polarization is caused by . . . , and is reduced by . . . .
5. The term "dry cell" is not strictly correct because . . . .
6. In a lead–acid cell the positive plate contains . . . , the negative plate contains . . . , and the electrolyte is . . . .
7. The characteristics of a lead–acid cell are: volts: fully charged . . . , discharged . . . ; specific gravity: charged . . . , discharged . . . .
8. An alternative secondary cell to the lead–acid type is the . . . , the two types of which are known as . . . and . . . .
9. The difference between the constant voltage and constant current charging circuits is . . . .
10. The current flowing when six $1 \cdot 5$ V cells each of $0 \cdot 75$ Ω internal resistance are connected in series to an external circuit of (a) $0 \cdot 25$ Ω, and (b) $32 \cdot 25$ Ω is: (a) . . . , (b) . . . .

Index

INDEX

Hammers   3
Hand tools   Topic No. 1
High temperatures   29–30
Hydrometer   99

Indicators, bell   89–92
Insulating   sheathed   wiring   Topic
No. 5
Intake position   47
Intermediate switches   67
Internal resistance of cells   102–5

Joint
boxes   77
box method   34

Knives   1–2

Lampholders   72–73
Lay of stranded cable   43
Lead–acid cells   98–110
Leclanché cells   96–98
Lighting
accessories   Topic No. 11
subcircuits   Topic No. 10
Looping-in method   34

Mercury oxide cell   98
Methods of wiring   34–35
MIAS cable   24
MICS cable   24

Nife cell   100

One-way switches   66

PILSA cable   24–25
Pliers   8–9
Plugs, connecting   60–61
Primary cells   96–98
PVC-insulated cable   22–23

Radial circuits   57
Regulations   19
Ring circuits   58–59

Saddles   35
Safety   Topic No. 2
Screwdrivers   3–6
Secondary cells   98–101
Sequence of intake equipment   50–51
Sheathed wiring   35–37
Socket outlets
and plugs   Topic No. 9
connecting   60
Spurs   62
Stranded conductors   43–44
Stripping cables   27–28
Subcircuits, lighting   Topic No. 10
Switches   66–68, 73–78
Systems of wiring   32–34

Terminals   30–31
Terminations   27–30
Three-plate ceiling roses   34–35
Transformers, bell   93–94
TRS cable   23
Two-way switches   67

Underground cables   24

Voltage ratings of cables   22
VRI cable   22

Wire gauges   44–45
Wiring
access   54
methods   34–35
systems   32–34
Wood chisels   9–10
Wood saws   7–9